China's Water Pollution Problems

Water pollution is one of the most serious problems plaguing China today, with millions of people drinking water unfit for consumption. The abysmal conditions of water pollution, water scarcity, and inadequate wastewater treatment are fuelling increasing social discontent, as people become more concerned by the scale and severity of the problem.

This book describes how and why China has ended up in such a dire situation, what the government is doing to address the problem, and the difficulties it encounters. The analysis is based on both grey literature (newspaper articles, NGO reports, Chinese government information) and on academic studies. The grey literature gives a voice to those who suffer from the pollution, their advocates, and government officers; it allows the reader to better grasp the conditions on the ground, and the impact of water pollution among people in different areas in China. The academic literature adds a theoretical perspective and brings these different case studies into a broader context.

This book will be of great interest to students of environmental pollution and contemporary Chinese studies looking for an introduction to the topic, as well as researchers looking for an analysis of China's environmental problems.

Claudio O. Delang is Assistant Professor in the Department of Geography of Hong Kong Baptist University.

China's Water Pollution Problems

Claudio O. Delang

Routledge
Taylor & Francis Group

LONDON AND NEW YORK

First published 2016
by Routledge
2 Park Square, Milton Park, Abingdon, Oxon OX14 4RN

and by Routledge
711 Third Avenue, New York, NY 10017

First issued in paperback 2017

Routledge is an imprint of the Taylor & Francis Group, an informa business

British Library Cataloguing in Publication Data
A catalogue record for this book is available from the British Library

Library of Congress Cataloging in Publication Data
Names: Delang, Claudio O.
Title: China's water pollution problems / Claudio O. Delang.
Description: Milton Park, Abingdon, Oxon : Routledge, 2016.
Identifiers: LCCN 2015050389 | ISBN 9781138669994 (hardback) | ISBN 9781315617879 (ebook)
Subjects: LCSH: Water--Pollution--China. | Water--Pollution--Government policy. | Water quality management--China. | Environmental policy--China. | China--Environmental conditions.
Classification: LCC TD301.A1 D45 2016 |
DDC 363.739/40951–dc23
LC record available at http://lccn.loc.gov/2015050389

ISBN 13: 978-0-8153-5583-0 (pbk)
ISBN 13: 978-1-138-66999-4 (hbk)

Typeset in Times New Roman
by Taylor & Francis Books

Contents

Illustrations

Figures

Tables

1 Introduction

Millions of Chinese people are drinking water unfit for human consumption. Many people have not drunk tap water in years, preferring instead to load up on bottled water, worried whether their water is safe to drink. As stated by Gleick (2009),

> China's water resources are over allocated, inefficiently used, and grossly polluted by human and industrial wastes to the point that vast stretches of rivers are dead and dying, lakes are cesspools of waste, groundwater aquifers are over-pumped and unsustainably consumed, uncounted species of aquatic life have been driven to extinction, and direct adverse impacts on human and ecosystem health are widespread and growing. [...] The major watersheds of the country all suffer severe pollution.
>
> p. 79

According to China Water Risk (chinawaterrisk.org), the availability of water in China has not only been taken for granted but is greatly overestimated, and problems of its scarcity are largely overlooked. China Water Risk reports that while being home to roughly 20 per cent of the world's population, China only has 7 per cent of the world's freshwater reserves. Water is required for agriculture, the generation of thermal power, washing of ores, and the production and manufacturing of metals, semiconductors, food, beverages, paper, chemicals, plastics, and much more. In particular the demand for agricultural products has increased tremendously, and at 61 per cent agriculture is the largest user of water. Moreover, 38 per cent of agricultural and 51 per cent of industrial output, amounting to CNY 46 trillion, is produced in water-scarce regions.

China Water Risk asserts that water scarcity and pollution are not only environmental risks but fundamental business risks and that the water crisis could threaten economic growth and social stability. Currently, "water resources are falling whilst demand for water rises" (China Water Risk, 2013). If the country continues to move forward at its present rate of water usage, experts project that the water supply will be unable to meet the water demand by 2030. China Water Risk (2011) contends that pollution exacerbates water scarcity: 19 per cent of the seven rivers and basins and 35 per cent of the 26 key lakes and reservoirs monitored are essentially useless for both agriculture and industrial use.

As with air pollution (Delang, 2016), the abysmal conditions of water pollution have fuelled increasing social discontent, which, in some cases, has culminated in violent outbreaks during protest demonstrations. Public awareness is rising as people become more concerned that the country needs to address the pressing issues of water pollution, water scarcity, and wastewater management. "Airpocalypse" and the shocking levels of air pollution burdening China may have captured all the headlines in the Chinese and international media, but just as dire and frightening as the smog-filled skies is the country's water pollution. Not only is there a crisis in regards to the severe and ongoing shortage of water, but also the deterioration of the quality of drinking water in China continues to be a problem that needs to be resolved.

This book seeks to give an overview of the conditions of water pollution in China, and the measures being taken by the government to solve the problem. The book is divided into six chapter. After this Introduction, Chapter 2 looks at the conditions of water pollution in China. I start by describing the different standards of water quality, which range from Level I (good quality for drinking) to Level V (severely polluted and unsuitable for any use), as set out by the Chinese government. I also review how the water is monitored and tested to ascertain the various levels of pollution, and describe the condition of China's water bodies, including the seven major river basins and the largest lakes. In Chapter 3, I look at the sources of industrial, agricultural, and domestic water pollution. I review the factors that have led to the gradual degradation of China's waters, and why water pollution reached such catastrophic levels that have caused many of its lakes and rivers to turn black, red, or milky-white, and toxic with pollution; and why thousands of dead pig carcasses were seen floating down the Huangpu River in 2013. This will lead to a discussion of the consequences of contaminated water (Chapter 4) on food safety and health, and the social, environmental, and economic costs. I also review people's growing awareness of the scale of the problem, and the resulting protests and acts of social activism to improve the conditions.

The volume ends with a discussion of possible solutions to combat China's water pollution (Chapter 5), what measures the government has taken, and its current plans of action. Finally, I conclude.

References

China Water Risk (2011) China Water Risk: A Portrait. Retrieved from http://chinawaterrisk.org/resources/analysis-reviews/china-water-portrait-past-future/

China Water Risk (2013) China Water Crisis. Retrieved from http://chinawaterrisk.org/big-picture

Delang, C.O. (2016) *China's Air Pollution Problems*. Abingdon, Oxon.: Routledge.

Gleick, P. (2009) The World's Water: 2008–2009. In Gleick, P. (Ed.) *The World's Water 2008–2009*. Washington, D. C.: Island Press.

2 The levels of water pollution

Water quality standards

Over the past three decades, China's rapid economic growth has led to the contamination and degradation of most rivers, lakes, reservoirs, and waterways. Poor sewage systems, industrial spills, and the extensive use of agricultural fertilisers and pesticides have been major contributors to the country's water pollution woes. Some experts contend that efforts to tackle China's water pollution problems will take longer, will be more expensive, and, ultimately, more difficult to fix than its air pollution problem.

In order to make sense of the issues surrounding China's water pollution crisis, first it is important to define what it is and briefly explore its origins. Hogan (2014) describes water pollution with the following words:

> The contamination of natural water bodies by chemical, physical, radioactive or pathogenic microbial substances. Adverse alteration of water quality presently produces large-scale illness and deaths, accounting for approximately 50 million deaths per year worldwide, most of these deaths occurring in Africa and Asia. In China, about 75 per cent of the population (or 1.1 billion people) are without access to unpolluted drinking water, according to China's own standards. Widespread consequences of water pollution upon ecosystems include species mortality, biodiversity reduction and loss of ecosystem services. Some consider that water pollution may occur from natural causes such as sedimentation from severe rainfall events; however, natural causes, including volcanic eruptions and algae blooms from natural causes constitute a minute amount of the instances of world water pollution. The most problematic of water pollutants are microbes that induce disease, since their sources may be construed as natural, but a preponderance of these

instances result from human intervention in the environment or human overpopulation phenomena.

Wang (1989) maintains that the problems of water pollution first occurred with the industrial development in the 1950s and became more serious in the early 1970s, as population growth and the expansion of industry and agriculture led to higher water consumption. Since that period, wastewater discharges increased tremendously, causing the country to ultimately encounter two areas of environmental concerns in water resource management: pollution and supply shortages. The author cites a survey examining 878 rivers in the early 1980s which indicates that 82 per cent of the waters were polluted to some degree: "More than 5 per cent of total river length had become fishless, while over twenty waterways were unusable for agricultural irrigation because of pollution" (p. 851). The report concludes with a prediction that the volume of wastewater would continue to increase and double by the end of the century.

Feng (2015) concurs with this report stating that China's 30 decades of transformation "from an impoverished farming-reliant country to the factory of the world" (p. 2) has imposed long-term environmental, social, and health-related costs. The researcher refers to studies showing the positive correlations between industrial growth and the growth in the total amount of discharged wastewater. Feng provides the following statistics:

> From 2001 to 2012, the total amount of discharged wastewater in China grew from 43.3 billion tons in 2001, to 68.5 billion tons in 2012. The country experienced a growth of 25.2 billion tons of total discharged wastewater in 12 years, with an average of 2.1 billion tons of additional discharged wastewater annually, at an average annual growth rate of 4.3 per cent. Rapid industrialization combined with inadequate regulation and lack of treatment facilities are among the main contributors to water pollution and environmental degradation.
>
> p. 2

Feng points out that the widespread dumping of toxic chemicals, agriculture runoffs, and industrial contaminated wastewater has poisoned most of China's water sources, both above-ground and underground. Yet, the official statistics may actually under-report pollution. According to Economy (2013), the frequency of testing at treatment plants is too low; only 40 per cent of the treatment plants in China's 35 major cities have the capacity to test for all 106 indicators which are supposed to have been assessed since 1 July, 2012 (see Chapter 5); there are only a few independent water-quality monitoring bureaus and most water testing is done by the same water treatment plants being evaluated; little information is provided from local governments as to the results of the tests; and "no water testing accounts for the contamination that occurs from the aging and degraded pipes through which the water is transmitted to Chinese households".

Rivers

A report released by the Ministry of Environmental Protection revealed that "80 per cent of urban Chinese rivers have been polluted, and many of them are foul-smelling and black all year" (Zheng, 2015). The report also disclosed that over 200 million tons of wastewater from industrial production and households had been discharged into urban rivers on a daily basis, and more than 60 per cent of the water shortages in the cities located in the southern regions was attributed to water pollution.

China has different standards of water quality based on the uses that can be made of the water. These go from Level I (pristine water typical of nature reserves) to Level V+ (useless water that may even be dangerous to touch) (Table 2.1). Figure 2.1 shows that Level I is only a small

Table 2.1 China's river pollution grades

Surface water quality	
I	Good quality for drinking, may need simple processing. More commonly found in pristine nature reserves.
II	Lightly polluted, but may be used for drinking after treatment.
III	Can be used for drinking after processing. Can be used for swimming and fishing.
Polluted	
IV	Can be used for industrial purposes or the service industry as well as recreational purposes if there is no direct human contact with the water (non-physical contact).
V	Can be used for irrigation (both for agriculture and general landscapes, such as urban parks).
Highly polluted	
V+	Severely polluted and unsuitable for any use.

Zheng, 2015

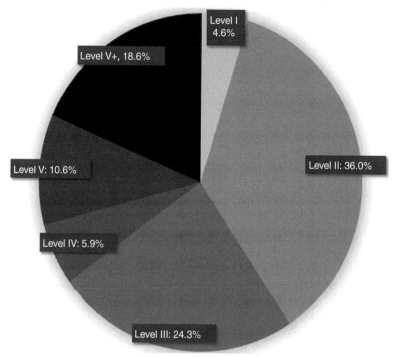

Figure 2.1 Quality of water in major river sections, 2014
Zheng, 2015

minority of the national waters (4.6 per cent of the total). This water is found in isolated areas where very little, or no human activities affect the quality of the water. Levels I–III account for 64.9 per cent of the river sections (Figure 2.1). This water may be used for recreational purposes or in households after treatment, or for fishing. Levels IV and V are severely polluted, and people should not come into contact with such water. Nationwide, 35.1 per cent of the river sections falls into this category. As I will be discussing below, such water is often used by farmers to water the crops, and in some cases, the pollutants are taken up by the plants, which may result in food poisoning.

Some of the most alarming statistics garnered from testing water are evidenced by the studies presented later in this book. According to Card (2014), in 2007 one-third of all Chinese rivers were polluted by agricultural, domestic or industrial discharge, while near urban areas the figure jumped to 90 per cent. Mengzhen Xu of Tsinghua University's State Key Laboratory of Hydroscience and Engineering

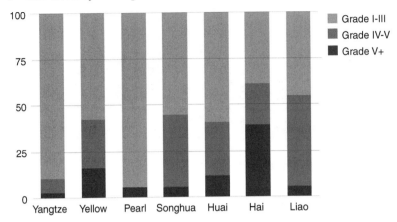

Figure 2.2 Water quality of China's seven main river basins, 2013
Du, 2015

remarked that the once clean Yangtze River that runs through nearly 4,000 miles of China's heartland is now, like many other rivers in the country, foul with contamination. He claims that most of the rivers are "either out of water or heavily polluted". Nevertheless, the situation is obviously not uniform throughout China. Figure 2.2 shows that the southern rivers, in particular the Yangtze and Pearl Rivers, are in much better condition because higher precipitation in the south means that the pollutants are able to dilute.

A 2004 survey (Shao et al., 2006) revealed that 28 per cent of the 412 monitored areas of seven major rivers in China were rated of no practical use. Yu et al. (2007) found that between 2001 and 2005, about 5 per cent of Chinese wells contained more than 50 µg/litre (micrograms per litre) of arsenic, which affected 582,769 people. Feng goes on to report that "30 per cent of industrial wastewater and more than 90 per cent of household sewage in China are released into rivers and lakes without treatment. Eighty per cent of a sample of 278 Chinese cities has no sewage treatment facilities, and 90 per cent of the cities with contaminated water have no plans of building any underground water supplies" (p. 2). Kaiman (2013a) uses similar statistics from China's Ministry of Water Resources, which state that in 2012 about 40 per cent of China's rivers were "seriously polluted", and that about "200 million rural Chinese have no access to clean drinking water". Zhou (2013a) also notes that of the 300 million people in China drinking contaminated water on a daily basis, 190 million drink water that is so contaminated it makes them sick. Furthermore, despite the fact that more than 10,000 petrochemical plants stretched along the Yangtze River, and 4,000 along the Yellow River, the Yangtze and the Yellow River supply drinking water for millions of people. According to Economy (2013), the Ministry of Supervision reported nearly 1,700 water pollution accidents annually. "The harbours of the Yangtze River and the Yellow River are already listed as dead zones by the United Nations – low oxygenated areas and de-oxygenated zones in which fish and shrimp cannot survive, and even sea plants cannot grow" (Zhou, 2013a, p.19). Unfortunately, these waters are not even considered the most polluted of China's seven major rivers.

Groundwater

In 2013, irate Chinese citizens posted their outrage on the microblogging site Sina Weibo in reaction to the rumours swirling around that factories in Weifang (Shandong province) continuously discharged wastewater into the region's aquifers – the principal source of drinking water for the city's 9 million residents. Officials confirmed it was true, further stoking people's fear that their drinking water was tainted (Li, 2013, p.14). According to Li (2013), 18 per cent of the water China uses comes from groundwater, and more than 400 of the country's roughly 655 cities have no other source of drinking water. "Much of the groundwater is contaminated, tainted by fertilisers, pesticide residues and dirty wastewater used for irrigation in China's vast rural regions, as well as pollutants from mining, the petrochemical industry, and domestic and industrial waste".

Data released in a 2012 report by the Ministry of Land and Resources disclosed that out of 4,929 groundwater monitoring sites in 198 prefecture-level administrative regions throughout the country, 41 per cent had poor water quality and 17 per cent extremely poor water quality, with levels of iron, manganese, fluoride, nitrites, ammonium and heavy metals exceeding safe limits. A government report in 2004 stated: "China had 38.8 million recorded cases of tooth-enamel damage owing to fluoride exposure; 2.84 million cases of bone disease owing to fluoride exposure; and 9,686 cases of arsenic poisoning" (Li, 2013, p.15). Yang Linsheng, the director of the Department of Environmental Geography and Health at the Institute of Geographic Sciences and Natural Resources shared with Li (2013), "these diseases are closely related to environmental and geological factors [and are] especially associated with contaminated groundwater" (p. 15).

Figure 2.3 shows that the conditions of underground water are even worse than those of surface water. Some polluting industries dump

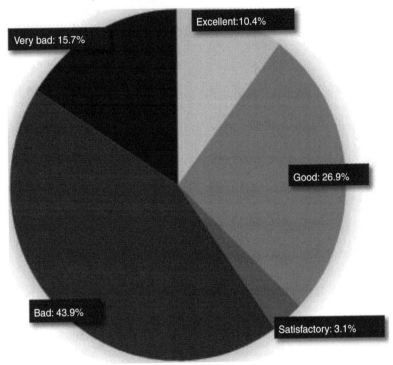

Figure 2.3 Overall water quality of national groundwater, 2013
China Water Risk, 2015a

their waste into underground water supplies, as do some municipalities. Because underground water does not recycle as rapidly as surface water, these pollutants remain in the water, resulting in higher pollution levels. The problem is that many cities, especially in the north of China, have to rely on groundwater for household consumption and agriculture. Du (2015) announced that China's Ministry of Environmental Protection discovered that among 4,778 testing sites in 203 cities, 44 per cent had "somewhat poor" underground water quality, and 15.7 per cent tested as "very poor". The water quality improved year-on-year at 647 sites, but worsened at 754 sites. According to China's groundwater standards, water of "somewhat poor" quality should only be used for drinking after it has been properly treated, while water of "very poor" quality should not be used as a source of drinking water.

Winsor (2015) cites a report by the Chinese Ministry of Environmental Protection, which states that nearly two-thirds of the groundwater (viewed as the worst offender) and one-third of the surface water was graded as unfit for direct human contact in 2014 and it is steadily becoming more polluted. The ministry "classified about 62 per cent of the 4,896 monitored groundwater sites as either relatively poor or very poor". Kaiman (2013a) quotes Ma Jun (Head of the Beijing-based Institute of Public and Environmental Affairs) as saying, "Groundwater is a key source of drinking water, industrial and agricultural use, especially in northern China, ... if this resource gets contaminated, it's far more difficult to restore than surface water or the air". Sun Ge, a research hydrologist, gloomily states, "Heavy metals are especially problematic, because once in the groundwater, they don't go away. It will be very expensive to clean up, if it is even possible" (Li, 2013).

Rodríguez-Lado et al. (2013) report that between 2001 and 2005, the Chinese National Survey Program conducted by the Chinese Ministry of Health tested some 445,000 wells in 20,517 villages of 292 counties (12 per cent of all counties in China) for arsenic contamination. In almost 5 per cent of the wells, arsenic levels were higher than the Chinese standard of 50 μg/litre. About 10,000 individuals were found to be affected by arsenic poisoning in known and suspected endemic areas (Yu, Sun and Zheng, 2007). The screening of the wells continued, and it has been estimated that up to 5.6 million people are exposed to high concentrations of arsenic in drinking water (>50 μg/litre) and that some 14.7 million are exposed to arsenic concentrations of >10 μg/litre (Amini et al., 2008). Rodríguez-Lado et al. (2013) concluded that "because of the size of China, it will take several decades to complete the screening of millions of wells to determine the spatial occurrence and magnitude of arsenic contamination throughout the country".

To make matters worse, a 2009 World Bank report found that "a deficit of surface water has led to excessive overexploitation of groundwater resources, which in turn has resulted in the rapid depletion of groundwater reserves" (China Water Risk, 2010a, p.6). The report outlines that the environmental consequences of groundwater depletion are lakes and wetlands gradually drying up, and an increase in the salinity of groundwater supplies, as well as soil subsidence, which causes damage to the infrastructure. Subsidence is explained as a process that "occurs when the aquifer is compacted from groundwater being depleted, which in turn causes the land above to subside and reduces storage capacity" (China Water Risk, 2010a, p.8). China Water Risk warns that if groundwater withdrawals are not managed properly and kept to sustainable levels, "China's economic progress will be threatened by water scarcity and the accompanying high cost of depending on a scarce and irreplaceable resource" (China Water Risk, 2010b, p.8).

Lakes

An investigative report by the official Xinhua News Agency referred to findings by the Chinese Ministry of Environmental Protection, and pointed out that, based on the standard set by the United Nations, 17 of China's 31 major freshwater lakes are polluted, and that "300 of China's 657 major cities also face water shortages" (Roberts, 2014). Kahn and Yardley's (2007) report that "in many parts of China, factories and farms dump waste into surface water with few repercussions. China's environmental monitors say that vast sections of China's great lakes, the Tai, Chao and Dianchi, have water rated Level 5, the most degraded level, rendering it unfit for industrial or agricultural use" (Figure 2.4). They also note that the World Health Organization found during this period that China suffered more deaths from water-related pollutants than from bad air.

Mengzhen Xu attributes the country's high pollution rates to the lack of awareness among the majority of the public, in addition to the limited restrictions on agriculture. He states, "Some people are sensitive to water pollution and hate the industries that destroy their homeland [...] however, some people don't care. They pour sewage water into the pools or rivers around their house and at the same time wash dishes in the same water" (Card, 2014).

Feng (2015) writes that approximately 10 per cent of China's surface water is heavily polluted and rated Level 5 or below, and that in some river zones up to 40 per cent of surface water is Level 5, while 90 per cent of the cities' underground water supplies are contaminated. These polluted waters not only damage nearby aquatic systems, but

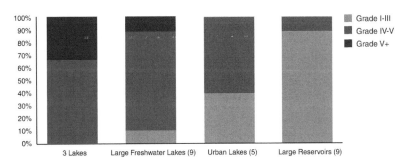

Figure 2.4 Water quality in China's 26 key lakes and reservoirs, 2011
Note: The three lakes are Tai Hu (in the Yangtze Delta plain near Shanghai), Dian Hu (in Yunnan), and Chao Hu (in Anhui Province)
China Water Risk, 2015a

flow into villages and towns, endangering the health of their residents, and contaminating underground water supplies. Feng reports that the World Bank even "warns China of the catastrophic consequences for future generations regarding its serious water shortages and water pollution problems" (p. 1). Feng refers to one study that shows that each day almost "980 million Chinese are drinking partly polluted water. Water pollution has become the main contributor to high rates of liver, stomach and oesophageal cancer in China" (p. 1). In correlation with this, the researcher mentions that a Beijing family made up of water expert officials revealed that they haven't drunk from the tap for 20 years. These officials observed that levels of nitrate in their drinking water rose from 1–2 mg/litre in 2007 to 9 mg/litre in 2013. The concern of finding hazardous contents in the water has spread nationwide.

Ocean pollution

Most discussions on the bodies of water affected by pollution are centred around the lakes and rivers. However, Zhou (2013a, p. 20) points out that China's oceans are also polluted and that although the seas contributed CNY2,160 billion to China's GDP in 2006, the rapid industrialisation increased the degradation of the ocean, making seafood unsafe to eat and damaging the local fishing industry. Zhou reported that untreated sewage is being regularly dumped into estuaries, thereby creating a surplus of red phytoplankton. The resulting red tides are killing off marine life and impacting coastal communities. Experts point to China's industries, agriculture, domestic sewage, oil and gas exploration, and fish farming as the reasons why its oceans are so polluted.

Moxley (2011) reports that 30 years of rapid economic growth has resulted in dangerous levels of ocean pollution. The State Oceanic Administration of China (SOA) reveals that 84,000 square kilometres of China's oceanic territory is viewed as seriously polluted, and that 14 of the 18 monitored ecological zones were considered to have unhealthy levels of pollution. In 2009, approximately 147,000 square kilometres of China's coastal waters failed to meet the standards for "clear water" due to coastal factories dumping an inordinate amount of industrial and domestic waste into the sea.

The Bulletin of Marine Environmental Status of China (SOA, 2010) reported that 86 per cent of China's estuaries, bays, wetlands, coral reefs, and seaweed beds were far below what the State Oceanic Administration (SOA) deems "healthy". State media also reported that oil, pesticides, and other harmful pollutants were contaminating China's marine life. The shellfish tested contained "excessive harmful chemicals," including lead, cadmium, and the insecticide DDT. Lead can cause damage to the human nervous system and blood and brain disorders if it is consumed in unsafe amounts – and the levels detected in shellfish were 50 per cent above normal.

Furthermore, Moxley (2011) disclosed that in 2010, China's coastal waters suffered 68 red tides or algae blooms, which were reportedly caused by excessive sewage in the water, affecting 14,700 square kilometres, 3.4 times the area affected in the 1990s. Yu Rencheng, a researcher at the Chinese Academy of Social Sciences' Institute of Oceanology, claimed that heavily polluted areas include the northern Yellow Sea, Liaodong Bay, Bohai Bay, Laizhou Bay, the Yangtze River estuary, Hangzhou Bay, and the Pearl River Estuary.

Monitoring water quality

The need to preserve and manage China's water resources has become urgent because of the country's rapidly depleting water quality. According to the 2014 Environmental Performance Index for wastewater treatment, which "tracks how well countries treat wastewater from households and industrial sources before releasing it back into the environment" (Miao, 2014), China ranks behind other emerging economies: China ranked 67th, behind Mexico (49th), South Africa (56th), and Russia (62nd). Wastewater treatment is even more important in China, because in many areas China has less precipitation to dilute the wastewater. This raises the question of how to monitor water quality.

Environmental Technology (2013) highlights some of the increasing number of water quality monitoring instruments that are being utilised in this drive to safeguard China's water resources, implying that the first requirement to achieve an effective management of the water resources is to obtain information and a thorough understanding of the situation. Rapid population growth, urbanisation, and industrialisation in the country threaten to limit the freshwater supply. In addition, strong seasonal variations in precipitation as a result of monsoons, and inadequate water and wastewater treatment facilities further compound these issues. It is noted that "one-third of China's main river systems have water of quality with very limited or no functional use, and in the water-scarce northern provinces 40 to 60 per cent are permanently classified as non-functional".

Environmental Technology (2013) contends that water quality monitoring networks are in place across Asia and that China maintains networks with thousands of monitoring sites. However, it is also noted that the spatial coverage remains sparse due to the size of the continent, and monitoring primarily consists of the "infrequent manual analysis of a few physical-chemical parameters suited neither to map out the dynamics of water quality viability nor to properly assess chemical and biological quality". As a result, China is investing in setting up networks of stations for ongoing water quality monitoring. In its 11th Five-Year Plan, China "intends to establish an advanced environmental monitoring system, equipping all key sources of pollution with automatic monitoring instruments" (Environmental Technology, 2013). This measurement of the basic physical-chemical parameters has been acclaimed as the most widespread application of on-site water quality instrumentation. Ion selective electrodes (ISEs) are being used for monitoring the elevated levels of nitrogen, mainly a result of fertiliser usage. Other technology and system devices for monitoring water quality include cabinet analysers used to detect contaminants such as metals and arsenic; ecological status monitoring to assess and inventorise the living organisms found in the water (i.e. green algae and/or cyanobacteria); toximeters or online biomonitors that continuously record an organism's behavioural and/or physiological response and evaluate any changes that could indicate pollution in the environment.

Water scarcity

Garland (2013) argues that water pollution represents only half of China's water problems – the other half being water scarcity. Garland conveys that "China supports 20 per cent of the world's population on only 6 per cent of the world's water" and that "freshwater reserves declined by 13 per cent between 2000 and 2009 alone". Of particular concern are 11 provinces in northern China, the so-called "Dry 11". These 11 provinces have a total year-long precipitation below the World Bank's "Water Poverty Mark" of 1,000 m^3; per person per year (Figure 2.5). The "Dry 11" are also highly industrialised, producing 51 per cent of the country's industrial output, and approximately 38 per cent of the country's food (China Water Risk, 2015b). Industries and agriculture are very polluting, so the low amounts of water mean that the pollutants cannot be diluted. This means that the little water available in the "Dry 11" also tends to be more polluted than the national average.

In a report on "China's Environmental Crisis", Xu (2014) writes that experts cite water depletion and pollution as China's biggest environmental hazards. Xu lists three variables: overuse, contamination, and waste as the primary factors that have produced the severe shortages. Furthermore, "approximately two-thirds of China's roughly 660 cities

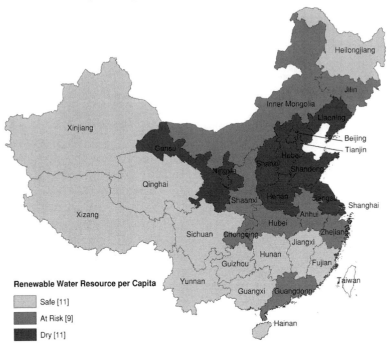

Figure 2.5 Uneven distribution of water resources in China
China Water Risk, 2015b

don't have enough water despite the fact that China controls the river water supply of thirteen neighbouring countries and has dammed every major river on the Tibetan plateau" (Xu, 2014).

Xu claims the impact is even more pronounced in the rural areas of the country, where about 300 to 500 million people lack access to piped water. In addition, Xu reports that negligent farming practices combined with the water crisis have turned some of China's arable land into desert, currently estimated at 27.5 per cent of the country's total land mass. Xu writes that according to the government "400 million Chinese lives are affected by desertification, and the World Bank estimates that the overall cost of water scarcity associated with pollution is around CNY 147 billion, or roughly 1 per cent of GDP".

The association between water scarcity and pollution is also discussed in a paper by Zhou (2013b). For people in metropolitan areas, "the shortage of water has resulted in a lack of drinking water" (p. 88). Zhou cites a survey indicating that out of the "more than 600 Chinese cities, two-thirds of them had inadequate water supplies, while one of every six experienced severe water shortages" (p. 88). Zhou goes on to say that as incomes increased and more Chinese people moved from older buildings into homes with modern plumbing, the amount of domestic wastewater drastically increased. Zhou notes that industrial pollution also contributes more to China's water shortages than had been previously assumed.

According to Garland, up to 24,000 villages "have been abandoned because of the desertification effects of the Gobi desert advancing eastwards". Moreover, Garland writes that chronic droughts in important agricultural regions complicate government efforts to provide food security:

> Beijing has long tried to maintain a balancing act between the spread of industry, continued support for agriculture and ensuring a clean supply for consumption by 1.3 billion people. As water supplies dwindle, competition may arise over which is given priority. All three are needed to ensure stability.
>
> Garland, 2013

Kahn and Yardley (2007) discuss that the severe water shortages could turn more farmland into desert. The authors say that while southern China remains relatively wet, the northern area, home to nearly half of the population, "is an immense, parched region that now threatens to become the world's biggest desert". Kahn and Yardley add, "Farmers in the north once used shovels to dig their wells. Now, many aquifers have been so depleted that some wells in Beijing and Hebei must extend more than half a mile before they reach fresh water. Industry and agriculture use nearly all of the flow of the Yellow River, before it reaches the Bohai Sea".

Ivanova (2013) posits that pollution is such a major component to water scarcity that it has been given its own special term: *shui zhi xing que shui*, meaning "water-quality-driven water shortage". China's pollution is more pronounced in areas where the economic surge is at its highest and water resources are most stressed. Official statistics reveal that "25 million of 295 million acres of cultivated land have been polluted, and more than 330,000 acres have been infiltrated or destroyed by solid wastes". Moreover, "half the soil in southern manufacturing cities is reportedly contaminated with cadmium, arsenic, mercury, petroleum, and organic matter" (Ivanova, 2013).

In a wide-ranging report on water pollution, Human Rights Watch (2011) proclaims: "Widespread pollution exacerbates water scarcity by compelling communities and factories to rely on contaminated water sources. As a result, water scarcity quickly becomes a public health problem". To elucidate this point, the authors look at how water scarcity in northern China forced farmers to irrigate approximately 40,000 kilometres with wastewater, causing crops and soil to be contaminated with heavy metal pollutants such as lead and mercury. This water scarcity "contributes to the spread of diseases associated with microbial and industrial pollutants". Human Rights Watch (2011) writes that industrial runoff and disasters profoundly impact water supply safety, especially since the release of toxic chemicals "can devastate an entire city". For example, after a chemical plant explosion incident in 2005, Harbin (the 10th largest city in China) was without water for its four million citizens. The water system was shut down for four days after the accident, but this didn't prevent "the release of 100,000 kg of benzene, aniline, and other heavy metals into the water system". This adds to another alarming statistic that "over 300 million people in China rely on hazardous water sources".

References

Amini, M., Abbaspour, K. C., Berg, M., Winkel, L., Hug, S. J., Hoehn, E. ... & Johnson, C. A. (2008) Statistical modeling of global geogenic arsenic contamination in groundwater. *Environmental Science and Technology*, 42 (10), 3669–3675.

Card, A. (2014, June 5) For Yangtze River and other polluted streams, Chinese scientists promote macroinvertebrate assessments. *Environmental Monitor.* Retrieved from http://www.fondriest.com/news/yangtze.htm

China Water Risk (2010a) China's water crisis: Part II – Water facts at a glance. Retrieved from http://chinawaterrisk.org/wp-content/uploads/2011/06/Chinas-Water-Crisis-Part-2.pdf

China Water Risk (2010b) China's water crisis: Part I – Introduction. Retrieved from http://chinawaterrisk.org/wp-content/uploads/2011/06/Chinas-Water-Crisis-Part-1.pdf

China Water Risk (2015a) Pollution status: Rivers, groundwater and lakes. Retrieved from http://chinawaterrisk.org/big-picture/pollution-status/

China Water Risk (2015b) Who's running dry? Provinces, autonomous regions and municipalities. Retrieved from http://chinawaterrisk.org/big-picture/whos-running-dry/

Du, Y. (2015, March 9) 9 Charts that show the level of pollution in China. CCTV America Newscasts. Retrieved from http://www.cctv-america.com/2015/03/09/9-charts-that-show-the-level-of-pollution-in-china

Economy, E. (2013, January 22) China's water pollution crisis. *The Diplomat.* Retrieved from http://thediplomat.com/2013/01/forget-air-pollution-chinas-has-a-water-problem/

Environmental Technology (2013, March 4) Water pollution in Asia – a brief review of monitoring technologies. Retrieved from http://www.envirotech-on

line.com/articles/water-wastewater/9/joep_van_den_broeke/water_pollution_in_asia_a_brief_review_of_monitoring_technologies/1384/

Feng, T. (2015, June 23) Water, the origin of life. Center for Sustainable Development. China Sustainability Project. Policy Brief Series. Retrieved from http://cgsd.columbia.edu/files/2015/06/PolicyBrief-WaterTheOriginOfLife-TingFeng.pdf

Garland, M. (2013, March 26) China's deadly water problem. *South China Morning Post*. Retrieved from http://www.scmp.com/comment/insight-opinion/article/1199574/chinas-deadly-water-problem

Hogan, C. (2014, November 17) Water pollution. The Encyclopedia of EARTH. Retrieved from http://www.eoearth.org/view/article/156920/

Human Rights Watch (2011) 'My children have been poisoned': A public health crisis in four Chinese provinces. Retrieved from https://www.hrw.org/report/2011/06/15/my-children-have-been-poisoned/public-health-crisis-four-chinese-provinces

Ivanova, N. (2013, January 18) Toxic water: across much of China, huge harvests irrigated with industrial and agricultural runoff. Circle of Blue. Retrieved from http://www.circleofblue.org/waternews/2013/world/toxic-water-across-much-of-china-huge-harvests-irrigated-with-industrial-and-agricultural-runoff/

Kahn, J. & Yardley, J. (2007, August 26) As China roars, pollution reaches deadly extremes. *The New York Times* (Asia Pacific). Retrieved from http://www.nytimes.com/2007/08/26/world/asia/26china.html?pagewanted=all

Kaiman, J. (2013a, February 21) Chinese environment official challenged to swim in polluted river. *The Guardian*. Retrieved from http://www.theguardian.com/environment/2013/feb/21/chinese-official-swim-polluted-river

Li, J. (2013) China gears up to tackle tainted water. *Nature*, 499(7456), 14–15.

Miao, W. (2014, April 16) What's behind the numbers in China's wastewater treatment plan. The Metri. Environmental Performance Index. Retrieved from http://epi.yale.edu/the-metric/whats-behind-numbers-chinas-wastewater-treatment-plan

Moxley, M. (2011, Jun 20) Pollution rising fast in China's seas. Global Information Network. Retrieved from http://www.ipsnews.net/2011/06/pollution-rising-fast-in-chinarsquos-seas/

Roberts, D. (2014, November 19) Think the air pollution is bad? China faces a water contamination crisis. *Bloomberg Businessweek's Asia*. Retrieved from http://www.bloomberg.com/bw/articles/2014-11-19/chinas-water-supply-is-contaminated-and-shrinking

Rodríguez-Lado, L., Sun, G., Berg, M., Zhang, Q., Xue, H., Zheng, Q. & Johnson, C. (2013) Groundwater arsenic contamination throughout China. *Science*, 341(6148), 866–868. doi: 10.1126/science.1237484

Shao, M., Tang, X., Zhang, Y. & Li, W. (2006) City clusters in China: air and surface water pollution. *Frontiers in Ecology and the Environment*, 4(7), 357.

SOA (2010) Bulletin of marine environmental status of China. Beijing: State Oceanic Administration of China (SOA). Retrieved from http://www.soa.gov.cn/zwgk/hygb/zghyhjzlgb/201211/t20121107_5527.html

Wang, J. (1989) Water pollution and water shortage problems in China. *Journal of Applied Ecology*, 26(3), 851–857. doi: 10.2307/2403696.

Winsor, M. (2015, June 04) China's pollution crisis: nearly two-thirds of underground water is graded unfit for human contact, report says. *International Business Times*. Retrieved from http://www.ibtimes.com/chinas-pollu tion-crisis-nearly-two-thirds-underground-water-graded-unfit-huma n-1953442

Xu, B. (2014, April 25) China's environmental crisis. Council on Foreign Relations. Retrieved from http://www.cfr.org/china/chinas-environmental-crisis/p12608

Yu, G., Sun, D. & Zheng, Y. (2007) Health effects of exposure to natural arsenic in groundwater and coal in China: an overview of occurrence. *Environmental Health Perspectives*, 115(4), 636–642.

Zheng, J. (2015, September 14) Public to have more say on pollution. *China Daily*. Retrieved from http://www.chinadailyasia.com/nation/2015-09/14/con tent_15316098.html

Zhou, J. (2013a) China's rise and environmental degradation: The way out. *International Journal of China Studies*, 4(1), 17–39.

Zhou, J. (2013b) *Chinese Vs. Western Perspectives: Understanding Contemporary China*. Lexington Books.

3 The sources of water pollution

Despite official statements that many of the problems concerning water pollution have been resolved, people are sceptical about what they are being told, comparing the official discourse to their own personal experiences. On March 14, 2014 the Ministry of Environmental Protection released a report stating that there were an estimated 280 million residents using unsafe drinking water in China. According to Lin (2014b), the situation is even worse in rural areas than in the cities, as approximately 110 million people in the vast rural areas lack access to clean drinking water. In the eyes of the public, this report, along with the slew of water pollution scandals that have been exposed and subjected to intense media scrutiny, has cast doubts on the government's ability to guarantee safe drinking water.

Essentially, there are a plethora of reasons for poor water quality in China, and since the pollution comes in different forms and is the result of a multitude of factors, there are no easy solutions.

First, pollution can be either point source pollution (pollution discharged through a pipe or some other discrete source) or nonpoint source pollution (pollution discharged over a wide land area, not from one specific location such as storm water runoff, agricultural drainage, and atmospheric deposition). The latter is more difficult to treat, because its collection is more technically challenging and expensive. This is especially true for agricultural pollution when the pollutants are introduced into the water over large areas, for example through the application of fertilisers or pesticides, or the excretion of free-range animals. Pollution from agriculture accounts for a large proportion of China's water pollution.

Second, pollution comes from a variety of sources. Apart from agriculture, it comes from industries and human activities. While most large cities have sewage treatment plants which treat most household waste, small conurbations usually have no means of treating household waste due to the high cost of building and operating the infrastructure.

Furthermore, water pollution is also caused by acid rain and air pollution.

Third, there is the problem of the diversity of pollutants in the water. Apart from sewage, there are artificial fertilisers and pesticides, heavy metals from industries, arsenic from mining, etc. As mentioned above, not all pollutants are measured, so in some cases the authorities responsible for cleaning the water do not even know what types of pollutants are present. This diversity of pollutants and sources makes treatment more difficult. The primary factors attributed to water pollution, coupled with examples of incidences, are examined in further detail in the following chapter.

Industry, dumping, and chemical spills

One of the largest polluters in China have been industrial plants such as chemical factories, drug manufacturers, makers of fertilisers, paper mills, and more. These factories are infamous for outright dumping poisonous chemicals and metals into the water used by residents for drinking. In other cases, the toxins have entered the water due to chemical and oil spills, leakage, or even factory explosions and mishaps. Examples of some of these incidents are highlighted below.

Sha (2014) disclosed that in early April 2014, excessive amounts of benzene – a toxic carcinogenic chemical used to manufacture plastics, lubricants, dyes, detergents, and pesticides – was found in the tap water system in Lanzhou (Gansu Province). According to Philips (2014), officials discovered 200 micrograms of benzene per litre, which is "20 times the acceptable national limit, in samples of the city's water supply". Philips states that benzene has been linked to leukaemia and non-Hodgkin's lymphoma. The author quotes the American Cancer Society warning that "consuming foods or fluids contaminated with high levels of benzene can cause vomiting, stomach irritation, dizziness, sleepiness, convulsions, and rapid heart rate. In extreme cases, inhaling or swallowing very high levels of benzene can be deadly".

Residents were cautioned not to drink tap water, and the provincial capital switched off its water supply for three days. Panicked residents rushed to stores to stock up on bottled water and eventually all the stores sold out the water, leading to citizens complaining they were dying of thirst. In some parts, the government supplied free drinking water, but the emergency water supply was cancelled as soon as authorities determined the tap water was safe – despite public scepticism and protest over their analysis, and the fact that there was a considerable delay before the authorities officially announced that the tap water had been contaminated.

Later, authorities announced that explosions that had occurred years before at a plant owned by oil giant China National Petroleum Corp (CNPC) may have been the cause of severe tap water contamination in the province. Their investigation led them to believe that the water source for a local water plant was poisoned by the oil pipeline leaks resulting from the explosions. Sha (2014) posits that "at least 34 tons of chemical residues were absorbed into the ground which then polluted an underground water duct opposite the site of the explosions".

In 2010, one of the largest recorded oil spills occurred along the Yellow Sea and off the shores of Dalian, which was once named "China's most liveable city" (Anna, 2010). Another in a series of environmental crises that is a result of the country's rapid industrialisation, one official warned that this spill posed a "severe threat" to marine life and water quality. According to Anna (2010) the spill originated from a pipeline owned by China National Petroleum Corp., Asia's biggest oil and gas producer by volume. The oil spread over 165 square miles of water five days after the pipeline at the busy north-eastern port exploded, "hurting oil shipments from part of China's strategic oil reserves to the rest of the country". One clean-up worker, a firefighter, died from drowning, his body covered in crude oil, when a wave threw him from the vessel. One Greenpeace worker said the crude was "as sticky as asphalt". Hundreds of officials, company workers, and volunteers worked on site and tirelessly cleaned up the blackened beaches with oil-skimming boats and fishing boats. An exasperated official from the Jinshitan Golden Beach Administration Committee told the Beijing Youth Daily newspaper, "We don't have proper oil clean-up materials, so our workers are wearing rubber gloves and using chopsticks".

A more recent oil spill occurred around the tourist island of Miaowan in late August 2015. Miaowan, located about 49 kilometres to the south of Hong Kong, is considered one of the most beautiful islands in Guangdong. Macau Daily (2015) reports the island was hit by heavy oil pollution, allegedly caused by a leak in a capsized vessel nearby. Information from the Guangdong Maritime Safety Administration conveys that roughly 200 square meters of sea near the pier of the island, dubbed "China's Maldives", was smothered in black oil which coated cargo boats and marine life. A task force dispatched by the mainland authority scrambled to remove the greasy pollutants from the sea.

Another case of a chemical spill was that in Changzhi (Shanxi province). Once again there was a delay in reporting the major environmental hazard to the residents. According to Wu, Sun and Peng (2013), a report in China Business News in Shanghai quoted a Handan official claiming that a chemical spill had occurred on 26 December 2012, five days before the leak was acknowledged by Changzhi authorities. The journalist reported that on 31 December 2012, plant employees conducting routine checks observed "aniline flowing out of the broken pipe. Approximately 39 metric tons leaked into an 80 sq. km river basin, and affected 28 villages in Shanxi and two other cities in the neighbouring provinces of Henan and Hebei". Aniline is a derivative of benzene used in industrial processes, such as dyeing and the production of pigments. "The reservoir was flooded by the toxic chemical which is believed to cause liver and kidney damage" (Wu, Sun and Peng, 2013).

Wu, Sun and Peng (2013) report that 5,000 people including soldiers, police, and workers armed with iron picks and shovels were involved in the emergency clean-up process. During clean up, temperatures dropped to −13°C, the upper layers of aniline froze, which allowed the clean-up crews to break the ice into pieces so it could be shipped away. The workers used activated carbon to absorb the chemicals that remained in the unfrozen water below. During the investigation, four high-ranking officials were terminated from the company, Tianji Coal Chemical Industry Group, which owned the plant – "a state-owned company that is one of China's largest producer of compound fertilisers". Although the website of the company claims that it was commended in 2011 for its achievements in environmental protection, energy conservation and emissions reductions, according to Wu, Sun and Peng (2013), in 2011 and 2012 the company was investigated and fined by the provincial environment department for the "excessive discharge of pollutants and irregular use of monitoring equipment".

Another case, this time of illegal chemical dumping, was described by Wei (2012). Wei (2012) reports that in March 2011, a procession of trucks began carting chemical waste from Luliang Chemical and illegally dumped it near the shores of Chachong Reservoir in Yunnan province. This went on for months and none of the residents were aware that the "yellow and black soil" carried by the trucks and later dumped by the roadside was poisoning the community's main water supply. Later calculations approximated that there were 5,000 tonnes of abandoned material. According to Wei, rainwater became contaminated after dropping on piles of poisonous chromium tailings, which then flowed into the 300,000 cubic meter reservoir, causing it to be toxic. Soon thereafter, 70 mountain goats and numerous farm pigs afflicted with swollen stomachs mysteriously died. Also, residents reported that the well water smelled strange and white shirts turned yellow after a washing. Consequently, "the three main crops in Yangqiying and Zhaishang villages were affected: tobacco leaves turned yellow and speckled, while rice and corn showed poor growth" (Wei, 2012). Soon the workers stopped drinking the well water and began carrying water from taps in the village. The villagers found out about the illegal dumping and attempted to stop the trucks, but to avoid the villagers the trucks came at night and continued their illegal activity.

The Environmental Protection Bureau for the local district of Qilin was alerted and began working on the clean up. Eventually, they were followed by the local government and a rainproof cover was placed over the waste piles and a ditch was dug to catch the polluted water. The site was patrolled, warning signs were displayed, and livestock was kept from being eaten or sold. The government later installed pumps and dug a channel to take the 300,000 cubic meters of water to the Nanpan River.

Authorities identified the hazardous waste dumped by Luliang Chemicals as "chromium tailings", which is described by Wei (2012) as a waste product from the manufacturing of chromium salts and metallic chromium. Tailings contain "1–2 per cent of calcium chromate (a carcinogen)" and "0.5–1 per cent of water-soluble hexavalent chromium (a deadly poison)". Luliang Chemicals claimed to be "Asia's largest producer of the chemical compound menadione, or vitamin K3. It is also one of Asia's largest manufacturers of chromium salts: each year, it produces tens of thousands of tonnes of them". The company was ordered to remove the contaminated soil and water to a safe place. During that process, 9,000 tonnes of tailings and soil were removed.

According to Wei (2012), a further investigation showed that a pond upstream of the reservoir, where goats drank from, had hexavalent chromium readings 2,000 times above normal levels. The toxin levels in Chachong Reservoir peaked at 200 times above standards. More excessive levels of chromium were also found at the entrance to the Huangnibao Reservoir downstream of Chachong. The residents of Xinglong village, where Luliang Chemicals is located, have always been aware of its dangers. Known by media reports as "a village slowly dying", 37 villagers were reported as having died from various forms of cancer since 2006. After an investigation, authorities claimed they were unable to find a link between the cancer cases and the chemical factory – they proposed that the cancers could have been caused by "bad dietary habits, such as eating fermented pickles, drinking alcohol or exposure to agricultural chemicals" (Wei, 2012). The only action that took place involving penalties was that the two truck drivers were arrested for the illegal dumping.

Gleick (2009) imparts that an epic disaster occurred on the Songhua River in 2005 when a chemical plant exploded in the city of Jilin and contaminated the river with 100 tons of benzene-related pollutants

(p. 83). The downstream flow of the contamination forced a temporary suspension of water supplies to almost 4 million people in Harbin, the capital of Heilongjiang Province. The contamination issue also affected the Russian city of Khabarovsk, which shares the Heilongjiang River with China. According to Tilt (2013, p. 1150–51), after examining a range of media reports on the 2005 benzene spill in the Songhua River, it was discovered that government authorities did not notify the public for over a week, but shut down the municipal water system of Harbin for several days under the guise that officials were conducting routine repairs. China Central Television investigated the cover-up, which resulted in the dismissal of MEP Minister Xie Zhenhua.

The Songhua disaster led to more ambitious efforts by the Chinese government to resolve their water quality issues. This led to plans by the government to build over 200 pollution control stations along the river at a cost of almost CNY16 billion. Gleick (2009, p. 83) writes that the city of Jilin (Jilin Province) built a new sewage treatment plant to "process a substantial amount of previously untreated waste". Notwithstanding, by September 2006, another 130 water pollution incidents had followed the Songhua River spill. According to Gleick, in 2007, local reservoirs around Changchun City in Jilin Province suffered a blue-green algal outbreak largely "attributed to pollutants from both industrial and agricultural sources, including both fish and pearl farms, which rely on the heavy use of fertilisers and pesticides. Such outbreaks led to the suffocation of native fisheries as the algae consume all of the oxygen in the water. The outbreak also threatened the quality of water, and led to a reduction in drinking water supply to the city of more than seven million inhabitants". Outbreaks were soon reported in other areas as well. During 2007, a series of water contamination incidents in Jiangsu Province in Eastern China resulted in the water supplies of millions of people being cut off.

On 31 December 2012, a spill by the Tianji Coal Chemical Industry Group, which uses water to convert coal to fertiliser at a factory in Changzhi City (Shanxi Province), affected at least 28 villages, a few cities such as Handan (in Hebei Province), and more than one million people. Wong (2013) writes that officials in Handan were irate when they learned they were not notified until five days later that a chemical spill had taken place at the fertiliser factory upstream. They had to immediately shut off the tap water, which sent the residents rushing to purchase bottled water; and they had to tell the farmers not to graze their livestock near the river. Xinhua, the state news agency, announced the results of the investigation on 20 February. The agency declared that a faulty hose caused the leakage of approximately 39 tons of aniline, a potential carcinogen, from the fertiliser factory. Eventually, 39 people were punished, including the mayor of Changzhi, who was removed from his post. Wong notes that Greenpeace East Asia issued its own report about the spill, elucidating that there were about 100 coal-to-chemical factories on the upper reaches of the Zhuozhang River, and that there had been a "history of clashes between the heavily water-consuming coal-to-chemical factories and the citizens downstream competing for water to drink". Greenpeace stated that Tianji is "notorious for its pollution". Larger factories like Tianji "use 2,000 to 3,000 tons of water per hour", which is "equivalent to the amount of water that more than 300,000 people use in a year" (Wong, 2013).

Domestic and industrial wastewater

Rising urbanisation also led to a rise in the production of domestic wastewater. The total discharge of wastewater increased by 65 per cent from 41.5 billion tonnes in 2000 to 68.5 billion tonnes in 2012, and is projected to continue rising with urbanisation and industrialisation. This total includes discharge from domestic use and industrial wastewater. A yearly discharge of 68.5 billion tonnes of wastewater is larger than the annual flow of the Yellow River of 58 billion tonnes (Hu, Tan, and Lazareva, 2014). Hu, Tan, and Lazareva argue that one crucial problem originates from the rapid development of the water supply sector in China, which ensured that the millions of people who moved to the cities over the last decades had water (Figure 3.1). This led to an

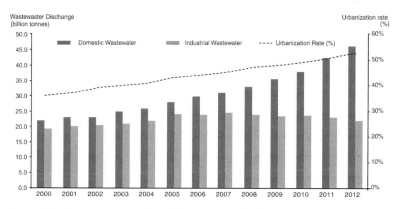

Figure 3.1 Domestic and industrial wastewater discharges vs. urbanisation rate (2000–12)
China Water Risk, 2014

increasing consumption of water, as water is more readily available to the urban population than it is to the rural one. However, the construction of sewage pipelines and, in particular, that of the wastewater treatment infrastructure did not keep pace with the increasing amounts of water used. The expansion of the wastewater pipe network and treatment facilities is included in the 12th Five-Year Plan, and the total investment in wastewater treatment and recycling infrastructure in urban regions is anticipated to reach CNY 430 billion. However, the target buildout of wastewater treatment plants for almost 3,000 cities lacking such facilities has been delayed.

Another question is related to the accuracy of the data regarding industrial wastewater. According to Hu, Tan, and Lazareva (2014), "industrial wastewater discharge appears to be under-reported". As Figure 3.2 shows, industrial output has grown continuously from 2000 to 2012, but official data show that industrial wastewater discharge has remained fairly constant, and even dropped from 2007 to 2012. During this period household wastewater has increased, and it is unlikely that industrial wastewater has dropped. If indeed the figures of industrial wastewater are not accurate, it is questionable how well the government can solve a problem that does not officially exist, and whether the government knows where the wastewater facilities have to be built.

The Hai River, which crosses five provinces, has been classified as "severely polluted" since the 1990s, and is viewed as one of the worst of all the polluted rivers in China. Yin (2015) cites a report by Tianjin Municipality, which states that 72.7 per cent of water in the Tianjin

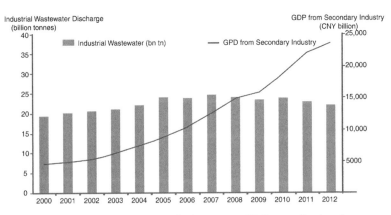

Figure 3.2 Industrial wastewater discharge vs. GDP contribution from secondary industry
China Water Risk, 2014

Municipality section is badly polluted. The region also suffers from severe water shortage. According to Bao Jingling, former chief environmental protection engineer in Tianjin, "The shortage of water has led to the accumulation of industrial, agricultural and household pollution in the river basin. The pollution is often washed into the city streets by torrential rain". Although the water quality has slightly improved over the last years, Jiangquan Wang, a professor at the School of Resources and Environmental Engineering under Hefei University of Technology in Anhui Province says, "Nearby tributaries and lakes have not received enough attention, so parts of the river are still in a poor state". Media reporters visiting a tributary of the river found it covered in white foam and dead weeds with a horrible stench. The local farmers remarked that the smell is especially foul in the summer and people who wade in the river develop a rash.

Sun, Chen, Chen, and Ji (2013) look at the reasons beyond wastewater for the pollution of the Haihe River, investigating total nitrogen (total N) concentration as a function of land-use patterns and comparing "the relative significance of the identified land-use variables for 26 upstream watersheds of the Haihe River basin" (p. 45). The authors argue that the growth of the human population and intense development over the past 30 years have profoundly impacted the ecological conditions in the Haihe River, which has upstream mountainous regions and downstream alluvial plains, and set out to investigate the impact of that development on nitrogen concentration. The authors conclude that the "total N concentration was 50 per cent human-induced land-use intensity, 23.13 per cent landscape patch-shape complexity, 14.38 per cent forest and grassland area, and 12.50 per cent landscape patch-area complexity" (p. 45). The researchers stress the importance of understanding the extent to which land-use patterns influence nutrient loading, because it is crucial to developing an effective water management system and a better decision-making process for improving water quality.

Groundwater also gets contaminated from industrial wastewater. One case is that of Weifang, a city of 8 million in the coastal province of Shandong known for its annual kite-flying festival. The city's internet users accused the local paper mills and chemical plants of routinely pumping industrial waste into the city's water supply 1,000 meters underground, causing cancer rates in the area to increase (Kaiman, 2013a). These allegations were disseminated through microblog posts by reporter Deng Fei who stressed, "I was just angry after receiving information from Web users saying that the groundwater in Shandong had been polluted and I forwarded it online. [...] But it came as a surprise to me that after I sent out these posts, many people from different places in northern and Eastern China all complained that their hometowns have been similarly polluted" (Liu, 2013). In a follow-up, Kamain reports that Weifang officials offered a reward of about CNY10,000 to anyone providing evidence of illegal wastewater dumping. Thereafter, local authorities investigated 715 companies without finding any evidence of wrongdoing.

Inefficient wastewater treatment and waste disposal

In some cases, pollution comes from wastewater treatment plants that do not treat the wastewater sufficiently. With the help of satellite images, Greenpeace (2014) identified a huge black plume of wastewater around "the size of 50 Olympic swimming pools" off the coast of South Eastern China next to the city of Shishi, a centre for children's clothing production. Following more extensive research, Greenpeace discovered the "plume" was emitted "out of a discharge pipe from the Wubao Dyeing Industrial zone and, more specifically, the Haitian Environmental Engineering Co. Ltd wastewater treatment plant, which serves 19 of Shishi's textile dyeing facilities". Greenpeace activists collected and tested samples of the water and found the presence of a multitude of hazardous chemicals, such as the hormone disruptor nonylphenol, chlorinated anilines, and antimony in the wastewaters.

According to Greenpeace, the toxic water pollution scandal uncovered at Wubao, Shishi City, is just one of many. In China, there are 435 discharge points like Wubao spanning the coast and releasing 32.2 billion tons of wastewater into the sea annually. Greenpeace cites official statistics from China's State Ocean Administration, which in 2012 pronounced that 68 per cent of discharge points had records for illegal discharge, and 25 per cent never met any national environmental standards.

Greenpeace also tested children's clothing purchased and produced in Shishi City along with another centre for children's textiles in the city of Zhili in Zhejiang Province. The two cities account for 40 per cent of all children's clothes made in China. The tests revealed that many of the chemicals found in the wastewater discharged from the dyeing facilities were also embedded in the clothing. The use of hazardous chemicals during the manufacturing of children's clothing presents a giant-scale problem, because not only it is leading to environmental pollution locally, but the residue from these substances can also be found in the products sold across China and exported to countries worldwide. Greenpeace adamantly stresses that the continued use of these hazardous chemicals not only in clothes, but also in children's toys and other products, will ultimately lead to increased levels being released into the environment.

> Given the scale of manufacture in the textile industry, the use of these chemicals, even at low levels, can lead to considerable amounts ending up in our environment, increasing children's exposure to these hazardous substances and heightening the potential health risks they pose. Compared to adults, children can be more sensitive to some effects of certain hazardous chemicals. Some chemicals have the ability to interfere with children's normal hormone functions and affect the development of the reproductive system, immune system or nervous system.
>
> Greenpeace, 2014

Greenpeace advocates that the Chinese government should enforce a policy that requires factories using and discharging hazardous chemicals to register and publicly disclose information on their discharge of hazardous chemicals. The organisation claims that currently there is no adequate regulation to oversee the use of hazardous substances used at the hundreds of production sites like those found in Shishi, and this is critical to making sure that hazardous chemicals are not used to manufacture clothing and other textiles for children and adults (Greenpeace, 2014).

Another case of poorly disposed waste is that of the Henan landfill, which threatened to pollute a reservoir that provides drinking water for more than 20 million Beijing residents (Luo, 2013c). In 1989, Xinglong County first began depositing rubbish in Qingsongling rubbish dump. Once a 50-meter deep valley, the rubbish dump was completely full when it was closed in 2009, occupying an area of 23,000 square meters. According to Luo, whenever heavy rains hit the region, floods carried garbage along a small river to the Yangzhuang reservoir, which connects to Jinhai Lake, Beijing's major drinking water source. The government's only response was to build a dam next to the former rubbish site designed to stop the rain from flushing refuse into the lower reaches, but the residents of Longwo, a nearby village, deemed it useless. Zhao Zhangyuan from the Chinese Research Academy of Environmental Sciences stated that valleys were not suitable sites for rubbish dumps and that the soil beneath was soft rock, allowing the pollutants to leak through and contaminate the groundwater. Residents of Longwo complain that each summer swarms of flies and bad smell plagued the area, in addition to the rotten pigs and chicken carcasses that attracted flocks of crows, which in turn severely damaged farm crops.

Similar problems exist for the water used for the upkeep of city parks. Larson (2014) cites a research study in Beijing in which researchers for the Chinese Academy of Sciences compared conditions in city parks watered with fresh water versus recycled water. The wastewater was inadequately treated before being used to water urban parks or redirected through scenic downtown canals. The researchers reported that "wastewater treatment plants are designed to remove solids, organic matter, and nutrients from water, but they are not properly equipped to treat the kinds of waste that may be found in used water from hospitals and pharmaceutical facilities". Wastewater plants don't remove traces of antibiotics and can even act as reservoirs for them, serving as a vector for spreading antibiotics. The study findings revealed that "urban parks in China doused with recycled water contained dangerously elevated levels of antibiotic-resistant genes, with quantities from 100 times to 8,655 times greater than in other parks" (Larson, 2014).

Agriculture and farming

Another major source of water pollution has been farmers' overuse of fertilisers and pesticides, which have contaminated the water and soil – as well as cast doubts on food safety. Over the last decades, Chinese farmers "have dramatically increased their use of chemical fertilisers, in part because they have become richer and can spend more on fertiliser, and also because of a rapid rise in the production of vegetables, which uses about 10 times more fertiliser than grain" (Rothman, 2006, p.11). According to the OECD, the average use of chemical fertilisers per hectare of sown land rose from 175 kg in 1990 to 289 kg in 2003, an increase by 65% over 13 years.

One of the reasons for the excessive use of fertilisers is that farmers use fertilisers to try to compensate for the little water available. Unfortunately, the biomass growth that is facilitated by fertilisers results in increased transpiration rates, which in turn further reduce soil moisture. This phenomenon has been studied by Liu, Pan et al. (2015) through *in-situ* measurements of soil moisture in agricultural plots in 40 stations in northern China. By analysing changes in soil moisture over 30 years, Liu, Pan, et al. found that topsoil (0–50 cm) volumetric water content during the growing season had declined significantly during 1983–2012. The team also conducted a long-term study from 1983 to 2009 at two contiguous sites, a pristine pasture and an agricultural site. The results showed that the soil moisture of the farmland dropped during the period under consideration while that of the pristine pasture increased, indicating that the drop in soil moisture of the farmland was due to farming practices rather than climatic

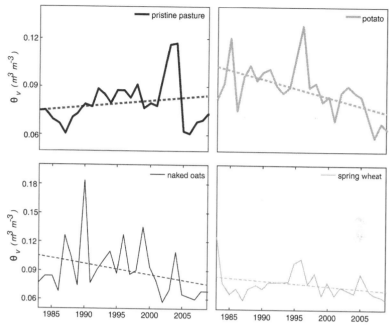

Figure 3.3 Variation of volumetric soil moisture (*θ*v) in topsoil (0–50 cm) of pristine pasture and different crop fields in Wuchuan Agricultural Meteorology Observation Station during 1983–2009
Source: Liu, Pan et al. (2015)

conditions (Figure 3.3). The reduction of soil moisture correlated with the increased fertiliser usage and the proliferation of crops with high water demands, like maize. The study notes that the use of fertiliser causes plants to grow larger and increases the number of leaves per plant. It also leads to increased water transpiration. "Fertiliser use may aggravate soil compaction and soil salinity, which reduces the water holding capacity of soil and, consequently, reduces available soil water", although they recognise that fertiliser use is not the only factor involved, and other agricultural practices may also play a part in drying the land (Gardner, 2015). This region in northern China only comprises 18 per cent of the country's water resources, but accounts for as much as 65 per cent of the arable land (Piao et al., 2010) and 40 per cent of the population of China. Figure 3.4 gives an idea of the extent of the problem, with some of the driest areas also being the most heavily farmed.

Study leader Qianlia Zhuang cautions that "If this trend continues, the soil may not be able to support crops by as early as 2090". Zhuang goes on to say "The soil moisture declined by 1.5 to 2.5 per cent every

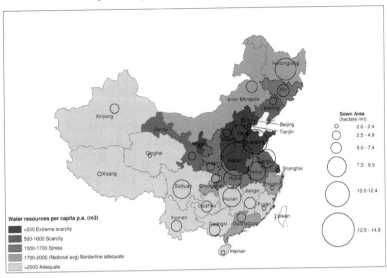

Figure 3.4 The state of China's agriculture
China Water Risk Review, 2014

decade of the study and while climate change is still a factor, this water depletion appears to be largely driven by human activities. [...] A 10 per cent decline in soil moisture over the course of a century would have major implications for agriculture and the freshwater supply in this heavily populated area" (Gardner, 2015). There are also considerable variations in soil moisture in different parts of the region, with some areas at the borders of the region experiencing an increase in soil moisture.

In another discussion on the use of fertilisers and pesticides, Meng (2012) adamantly states that their overuse is causing pollution and food safety problems, and that agriculture in China has become reliant on those chemicals. Meng cites figures from the Chinese Academy of Agricultural Sciences' (CAAS) Soil and Fertiliser Institute which show that in half of all Chinese regions, the average levels of nitrogen fertiliser use exceeds the internationally accepted limit of 225 kg per hectare. Ironically, only 30 per cent of the fertiliser applied is actually used by the plants to grow. In Henan Province, three million tons of fertiliser are applied by farmers, but only one million tons are absorbed by the crops. It is mostly the portion of unused fertiliser which causes the pollution.

A Chinese Academy of Agricultural Sciences (CAAS) "survey of intensive vegetable farms in 20 counties in five northern provinces found that half of 800 points surveyed had excessive levels of nitrates in groundwater attributable to fertilisers" (Meng, 2012). A researcher at CAAS's Institute of Agricultural Resources, Huang Hongxiang, said that planting year-round requires the large-scale use of pesticides and fertiliser, and this inevitably creates soil and water pollution, which in turn means even more fertiliser and pesticides are needed, creating a vicious cycle. Zhang Yang of the Central Rural Work Leading Group Office stressed that "currently China's agricultural and rural ecosystems are vulnerable. Soil and water are being lost, the land is degrading, crop diversity is falling, natural disasters are frequent, and the excessive and inappropriate use of fertilisers and pesticides mean that both farms and villages are badly polluted. Agricultural and rural pollution will cause a range of problems, including with food security" (Meng, 2012).

Wang, Ma, Fan, Zhang and Qian (2014) look at the water quality of Dianshan Lake in Shanghai Municipality, and in particular the impact of agricultural lands around the lake. Dianshan Lake is the largest freshwater lake in Shanghai Municipality and sits in the Taihu Lake Basin and the Yangtze River Delta. Wang, Ma, Fan, Zhang and Qian monitored nutrient losses from different types of agricultural land use and variations of nutrient concentration. In particular, they analysed the relationship between the nutrient losses from agricultural nonpoint sources (NPS) and nutrient stocks in Dianshan Lake over monthly and seasonal time periods. The main findings were that "the monthly average concentration of total nitrogen (TN) ranged from 1.41 to 7.34 mg/L in 2010 and from 1.52 to 5.90 mg/L in 2011, while the monthly average concentration of total phosphorous (TP) ranged from 0.11 to 0.26 mg/L in 2010 and from 0.13 to 0.30 mg/L in 2011. The annual loss of TN from agricultural NPS was of 195.55 tons in 2010 and 208.40 tons in 2011". Accordingly, the monthly TN and TP losses from agricultural NPS showed a positive correlation with the monthly TN and TP stocks in the lake in 2010 and 2011 (pp. 381–382). The authors suggest that the main focus of future studies should be on controlling the pollution caused by local agricultural activities. These recommendations are echoed by Liu, Ju et al., (2006), who state that "as the largest developing country in the world, China has consumed more than 24 Tg per year fertiliser N in recent years. This is about 30 per cent of total fertiliser N used worldwide" (p. 370–1).

Another example is that of Erhai Lake, an iconic lake once famed for its crystal clear waters in the south-western province of Yunnan, which has turned milky-white with pollution (Radio Free Asia, 2014a). Residents posting infamous pictures online of the polluted lake, and reports by social media sparked an investigation by the Dali prefectural environmental protection department, which was pressured to

implement clean-up measures and to pursue and punish those enter-
prises suspected of causing the pollution. Many experts claim the lake's
pollution problem was caused by "unauthorized waste outlets". Some
residents blamed the pollution on the influx of migrants from other
parts of China, while others said there were few companies on the lake
and blamed the government for not keeping up with the rapid rise in
waste production. Activists claimed that cleaning China's highly polluted lakes and rivers overall could take longer that the three decades it
took to pollute them (Radio Free Asia, 2014a).

Zhao, Chen, and Zhang (2013) describe nitrate as one of the most
common inorganic groundwater contaminants in the world. It can
cause methaemoglobinaemia in infants and even in the stomach of
adults if highly concentrated in drinking water. Ammonia in drinking
water is not regarded as health related, yet it is an indicator of possible
bacterial sewage and animal waste pollution. A high enough concentration can potentially increase the nitrate concentration through
nitrification. Zhao, Chen, and Zhang conducted a study to identify the
current status of nitrate and ammonia contamination in the drinking
water in wells in Hailun (Heilongjiang Province), a city made up of
23 townships and a large area of farmland and forestland, and an
important base for grain production. Zhao, Chen, and Zhang set out
to identify the factors affecting nitrate and ammonia contamination in
drinking water in rural areas, and to offer suggestions to ensure
drinking water safety in China's rural areas. The authors found that
"32.89 per cent of the domestic wells were contaminated above the
maximum acceptable concentration (MAC) of nitrate and 21.05 per
cent were contaminated above the maximum ensured concentration
(MEC) of ammonia" (Zhao, Chen, and Zhang, 2013, p.31). In line
with other studies (Bonton et al., 2010; Thorburn, Biggs, Weier and
Keating, 2003), Zhao, Chen, and Zhang found that chemical fertilisers
were the major source of nitrogen contamination in the groundwater.
To ensure the safety of the drinking water, they recommend reducing
the inputs or improving the efficiency of chemical fertilisers.

The Lake Tai

Lake Tai has been the subject of much research because of its cultural,
ecological and socioeconomic importance. The case of the Lake Tai
(also called Taihu, Figure 3.5) provides an example of pollution in a
large (2,250 km^2) freshwater lake. Lake Tai is the source of the Suzhou
River, which feeds the Huangpu River that passes along Shanghai's
Bund, recognised by many as the country's most famous riverfront.

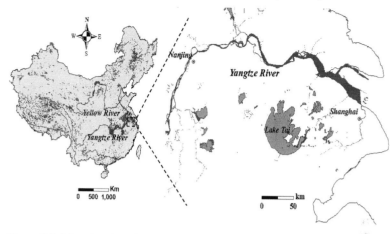

Figure 3.5 Map showing the location of Lake Tai
Huang, J. et al., 2015

The water channels that criss-cross the region around Lake Tai have been viewed as the arteries of the region's economic development. "It is no exaggeration to say the lake is the heart that drives the seven cities of Shanghai, Suzhou, Wuxi, Changzhou, Hangzhou, Jiaxing, and Huzhou" (Liang and He, 2012).

Renowned at one time for its natural beauty and for being the hub of a land rich with fish and rice, Lake Tai has suffered greatly from the decades of major pollution problems. In 2007, Lake Tai, the third largest freshwater lake and one of the most important water resources in China, was beset by toxic cyanobacteria, commonly referred to as pond scum, which caused the huge lake to become fluorescent green and emit a foul stench that drifted over neighbouring villages (Kahn, 2007). Jun (2007) describes how a fetid bloom of blue-green algae caused the water quality in the city of Wuxi, in East China's Jiangsu Province, to severely deteriorate. The water became putrid and very unpleasant to drink. Kahn reports that nearly two million people living amid the canals, rice paddies, and chemical plants surrounding the lake were forced to stop drinking or cooking with their main source of water.

Wu Lihong, who grew up by Lake Taihu, remembers what it was like years ago: "Twenty years ago, the water was clear. I used to swim in it as a child" (quoted in Liang and He, 2012). Wu can recall streams teeming with whitefish tickling his legs. Before the area's downturn, affluent residents built lush gardens that featured the lake's "wrinkled, water-scarred limestone rocks set in groves of bamboo and chrysanthemum".

The region's environmental degradation began in the 1950s with the construction of dams and weirs to improve irrigation and control floods, which disrupted the cleansing circulation of fresh water. With strong local government support, chemical factories proliferated in the region throughout the 80s. Eventually, the northern arc of Lake Tai was saddled with 2,800 chemical plants consuming and discharging enormous quantities of water into the lake. According to Huang Xuanwei, formerly chief engineer at the Taihu Basin Agency (Liang and He, 2012) in 1987, around 36 tons of wastewater (half from Shanghai) were dumped into the area's lakes every year – 80 per cent was left untreated. Wu recollects that around 1988, the local fishermen found that the fish they caught had a strange taste. The rivers around Lake Tai gradually began changing colours – some turning into milky-white, others black. In the early 1980s, according to a pan-regional evaluation of water quality, 40 per cent of the rivers of the Taihu Basin was polluted; by 1996, this figure rose to 86 per cent – higher than any of the other Chinese water systems. Organic pollution of the severely damaged lake was reported to have risen from 1 per cent in 1987 to as high as 29.18 per cent by 1994.

By the early 1990s all the fish disappeared. Liang and He (2012) explain that "from 1993, the entire lake suffered eutrophication, the process where an excess of nutrients such as nitrates and phosphates causes an algal bloom that removes oxygen and reduces water quality, causing fish and other organisms to die". Over time the algal blooms became increasingly more severe and that is when the crisis happened. In 2007, a bloom described by Liang and He as "dozens of centimetres thick" covered the entire lake. Figures presented by the Wuxi government indicated that plants supplying at least 70 per cent of the city's water were affected – evidenced by the polluted tap water that served as drinking water to two million Wuxi residents turning yellow and foul.

In 2001, Wen Jiabao, then a vice premier and later China's prime minister, arrived to investigate reports of Lake Tai's deterioration. As usual, word of the Communist Party inspection tour was received in advance by local officials. When Wen Jiabao asked to see one of the typical dye plants, one had already been readied according to people who witnessed the preparations: "The factory got a fresh coat of paint. The canal that ran beside it was drained, dredged and refilled with fresh water. Shortly before Mr. Wen's motorcade arrived, workers dumped thousands of carp into the canal. Farmers were positioned along the banks holding fishing rods. Mr. Wen spent 20 minutes there. A picture of him shaking hands with the factory boss hangs in its lobby" (Liang and He, 2012).

The Economist (2010) reported that the government was shamed by protests and demonstrations, and spent hundreds of millions of dollars on sewage works and other measures. Even the most dedicated efforts still made it a struggle to keep the lakes clean and progress has been slow. Slops of oily green algal film are still found on the shores of Lake Tai and foul odours still waft throughout the surrounding villages, and fishermen still complain of dizziness. Regardless of the government's efforts to stem the tide and clean it up, the algal blooms still occur every year.

According to Liu, Liu, Zhang, and Bi (2013), in the past three decades "eutrophication caused by surplus nitrogen (N) and phosphorus (P) has become one of the most rapidly growing environmental crises of surface water". The dramatic increase of nutrient inputs have degraded water quality in many rivers and lakes. Liu, Liu, Zhang, and Bi focus on the Tai Lake Basin, which has been plagued by major eutrophication and is considered one of the main targets of water pollution control. Liu, Liu, Zhang, and Bi developed an empirical framework to estimate nutrient release from five major sectors: industrial manufacturing, livestock breeding (industrial and family scale), crop agriculture, household consumption (urban and rural), and atmospheric deposition. The results showed that among the five major sources addressed, "the household consumption sector was found to be the major contributor with the greatest impact on surface water (46 per cent in N load and 47 per cent in P load), whereas household wastewater discharge was the major emission source. Atmospheric deposition and animal excreta loss from livestock farms also contributed a significant share of nitrogen and phosphorus, respectively" (Liu, Liu, Zhang, and Bi, 2013, p. 735).

A different study based on Lake Taihu's degradation examined the correlations between algae and water quality, and the key driving factors for the lake's eutrophication. Li, Tang, Yu, and Acharya (2014) set out to investigate the spatiotemporal distributions of correlations between chlorophyll-a and water quality indices, "in order to clarify the main principal components from all the water quality parameters and identify the key driving factors for the lake eutrophication" (p. 171) due to natural or anthropogenic influences, and characterise the spatiotemporal distribution of those parameters. The analysis concluded that water temperature played an important controlling factor for algal growth in the lake. The study identified eutrophication as one of the key water quality problems for the lake, and concluded that nutrient contamination from anthropogenic and natural inputs was the principal environmental issue of the waters. In addition, the researchers write: "Lake Taihu is characterized with high spatial and seasonal variations in water quality and eutrophication, decreasing from north to the centre and from west to east, suggesting that algal bloom events occurred only in sections of the lake rather than the entire lake" (Li, Tang, Yu, and Acharya, 2014, p. 180). Moreover, the most heavily polluted areas in the northern and north-western sections of the lake strongly correlated with the major adjoining rivers in the most densely populated and industrialised regions, demonstrating the influence of watershed pollutant loads on the quality of the lake water.

One of the pollutants that has not been discussed in relation to Lake Taihu is petroleum. Guo, Fang, and Cao (2012) describe petroleum as "a complex mixture mainly composed of hydrocarbons, in which most of the alkanes are proved to be narcotic and irritant, and most PAHs (polycyclic aromatic hydrocarbons) have strong toxicity, carcinogenicity, teratogenicity and mutagenicity". They go on to say that "large amounts of oil pollutants will cause serious pollution to the water ecosystems and direct harm to the health of all living creatures". The study concluded that due to the rapid economic development of the Lake Taihu area, "the petroleum products and oil wastewater produced in the process of oil processing, transportation and application of various refined oil have been inevitably discharged into Taihu Lake" (Figure 3.6) (Guo, Fang, and Cao, 2012, p.1). In terms of improving the water

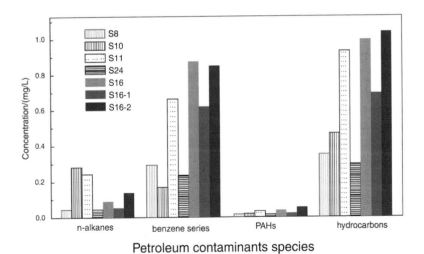

Figure 3.6 The concentrations of petroleum-type pollutants in the water samples collected in different regions
Guo, Fang and Cao, 2012

quality of Taihu Lake, Guo et al. (2012) recommend scientific methods for treating oily wastewater; a stronger management of lake pollution; a greater focus on the control over pollution sources, such as industrial and household sewage and oily wastewater from vessels; and the dredging of sediments in the most polluted locations.

Acid rain

Acid rain is rain with a pH level of less than 5.6 (neutral is 7). Acid rain is caused by the excessive emission of sulphur dioxide (SO_2) and nitrogen oxides (NO_x) mainly from the burning of fossil fuels. It causes the acidification of lakes and streams and damages trees and many sensitive forest soils, and even buildings. Prior to falling to earth, sulphur dioxide (SO_2) and its particulate matter derivatives harm public health. Given China's reliance on coal, it is not surprising that acid rain is a problem in many areas. In 2005, a report found that 28 per cent of the country's territory, mostly south of the Yangtze River, was affected by acid rain (NDTV, 2011). In 2009, official statistics showed that 258 Chinese cities experienced acid rain, while "every drop of rain in Xiamen in the first half of 2010 was acidic" (NDTV, 2011). The situation fluctuates yearly, and over the last few years there has been an improving trend. In 2015, out of the 470 cities nationwide that measure the pH of precipitation, 164 were still affected by acid rain. Nevertheless, during the first half of 2015, acid rain cities with an average pH value of rainfall lower than 5.6 dropped by 4.5 per cent year-on-year (Xinhuanet, 2015).

Heavy metals

Cheng, Wang, Wang, Wang, and Zhao (2012) looked at the degree and sources of heavy metal pollution in the Yellow River Wetland Nature Reserve of Zhengzhou. The Yellow River, which is the second longest river in China, has become infamous in recent years for the heavy metal pollution in the sediments due mainly to crude oil consumption and coal combustion. Cheng et al. explain that sediments are the "main sink" for pollutants and are viewed as significant indicators of water contamination (i.e. metals discharged). "In the aquatic environment, heavy metals tend to be incorporated into the bottom sediments and can be released by various processes under favourable conditions. Thereby metals reach the aquatic life community and human beings through the food chain and cause great concern because of the potential health risks to the local inhabitants" (Cheng et al., 2012, p.26).

Ren, Zhao, Sun, and Zhong (2013) set out to analyse the amount of heavy metals in the Tonghui River. The four metals focused upon in the study are lead, cadmium, copper, and nickel. The researchers describe the dangers of each of these metals to humans and animals: lead (Pb) is highly toxic to animals and humans due to its adverse effects on the hematopoietic, nervous, renal, and skeletal systems; cadmium (Cd) is a toxic element that is known to damage organs such as the kidneys, liver, and lungs; copper (Cu) in excessive amounts can cause nausea, vomiting, haemolytic jaundice, kidney failure, and depression of the central nervous system; and nickel (Ni) in low concentrations can cause an allergic reaction, while certain nickel compounds may be carcinogenic. The researchers detail how the extensive use of heavy metals in a variety of industrial processes and products resulted in widespread environmental contamination, and identify heavy metal contamination as one of the main sources of water pollution. They explain that "Heavy metals from the aquatic environment may accumulate in the human body through the food chain, causing serious health effects, including reduced growth and development, organ damage, nervous system damage, cancer, and death. Heavy metals are known to damage human health and to cause serious toxicity, being difficult to metabolize and easily bioaccumulated" (p. 1752). Fifty-five samples were collected from the Tonghui River in Beijing to determine the levels of heavy metals. The results of their analysis indicated that the levels of Pb and Ni showed increasing trends along the river from upstream to downstream, possibly attributable to the higher distribution of chemical factories in the downstream areas.

Xiao et al. (2013) analysed six heavy metals (Cd, Cr, Cu, Ni, Pb and Zn) in all sediment samples collected from the upper, middle and lower reaches of both urban and rural rivers in a typical urbanisation zone of the Pearl River Delta, in the south of Guangdong Province. They also evaluated spatial distribution, pollution levels, toxicity and ecological risk levels. Their study "concentrates on the sediments from two tributaries of the Pearl River watershed flowing through the Panyu district. One tributary (the Shiqiao River, flowing through the most developed urban area) has been polluted by domestic sewage, industrial effluent and waste leachate from primary electroplating, dyeing, and leather production since the 1980s. The other tributary (the Shawan River, flowing through a less developed rural area) is contaminated by agricultural effluent and wastes from such locations as food plants, grease and meat plants, chemical plants, heavy machinery manufacturers, hardware plants and shipyards" (p. 1565). Their results show that, among these six heavy metals (Cd, Cr, Cu, Ni, Pb, and Zn), Cd pollution was the most serious, especially in sediments from the upper reach of the urban river and the middle reach of the rural river. Both Cu and Zn exhibited higher levels of enrichment, whereas higher toxicity of Cr and Ni were observed in surface sediments. Moreover, the typical vertical distributions of heavy metals, SOM (soil organic matter), and grain size along typical sediment profiles from the urban and rural rivers suggested that dredging projects, SOM content, and grain size have important impacts on the distribution of heavy metals in sediments. The moderate to serious pollution levels, eco-toxicity, and ecological risks of these heavy metals in both rivers suggested the presence of widely dispersed pollution sources within and around the city of Panyu. The "hot area" with higher TEF and RI values in the upper and middle reaches of the urban river may demonstrate nonpoint source pollution in this densely populated area. However, a specific "hot spot" in the middle stream of the rural river may reflect point source pollution in the central rural area.

p. 1573

China's water-energy nexus

After irrigation, the energy sector is the second largest water user in the world in terms of water withdrawal (Hightower and Pierce, 2008). Almost every stage in the energy supply chain needs large amounts of water (Mielke, Anadon and Narayanamurti, 2010). This includes the drilling and fracturing in oil and gas exploration (Hu and Xu, 2013); the cooling or processing of thermal power generation (Liu and Wen, 2012); and the production of feedstock and biofuels (Zhang, Xie and Huang, 2014). Besides the impact in terms of the magnitude of water consumed and withdrawn, the production of energy also causes serious degradation to the local aquatic environment. For example, fossil fuel extraction, especially coal mining and shale gas drilling, generates considerable pollution, which often affects aquifers and leads to serious damage to the ecosystem and health (Greenpeace, 2012; Vidic et al., 2013).

Given the expected growth of the Chinese economy and the correlation between economic growth and energy consumption, the water scarcity that exists in China is particularly worrisome. In some parts of the world, the generation of electricity has already been affected by water shortages caused by the overexploitation of freshwater resources (Hightower and Pierce, 2008). In the future, China can expect to meet increasing challenges. First, because climate change will likely change the regional distribution of precipitation with increasing drought in some areas and precipitation in other areas, therefore introducing considerable uncertainties (Chandel et al., 2011). Second, because the Chinese economy is expected to continue growing, as is its population, the demand for energy is also expected to grow. China is already a major energy producer, accounting for 17.3 per cent of the global energy production and 17.5 per cent of the global consumption in 2010 (IEA, 2012). Davies et al. (2013) estimate that the global use of water for electric power generation will increase by 4–5 times between 2005 and 2095.

The question of the impact of energy production on water in China has been addressed by Zhang and Anadon (2013) who look at the life cycle water use of energy production and its environmental impact in China. They quantify "life cycle freshwater withdrawals, consumptive water use, and wastewater discharge of eight energy products (namely, coal, crude oil, natural gas, petroleum products, coke, electricity, heat, and gases) at the provincial level" (p. 14460). The definition of consumptive water use adopted in the study is consistent with that of a blue water footprint (Hoekstra et al. 2011).

Zhang and Anadon's (2013) findings are summarised in Table 3.1. Electricity is found to have the highest water use intensity across all indicators. The life cycle water withdrawals intensity of electricity is of 5,263 m^3/TJ, the life cycle water consumption intensities is 234.2 m^3/TJ, and the life cycle wastewater discharge intensities of electricity is of 656.7 m^3/TJ (or 18.9 m^3, 0.84, and 2.36 m^3/MWh). (Table 3.1). Direct water use accounts for 82 per cent of the electricity life cycle water withdrawal, 63 per cent of life cycle water consumption, and 39 per cent of life cycle wastewater discharge. The life cycle water use of all other energy products is much lower than that of electricity. For the other seven energy products, a large proportion of their life cycle water use is used upstream in the supply chains.

It requires a lot of water to produce electricity using coal, and this water becomes heavily polluted (coal has to be washed, etc.). Since there is little water in China, the need for much water to produce electricity using coal is worrisome. The life cycle water withdrawals, water consumption, and wastewater discharge intensity of coal are 106.4 m^3/TJ, 41.5 m^3/TJ, and 38.4 m^3/TJ, respectively (or 2.22 m^3/ton, 0.87 m^3/ton, and 0.8 m^3/ton), in which 17 per cent, 22 per cent, and 74 per cent are direct water use. Zhang and Anadon (2013) note that "coal is the only energy product whose direct wastewater discharge intensity exceeds its direct water withdrawal intensity. This is because large volumes of mine drainage are generated and discharged during the coal mining process" (p. 14461). The water intensity of the "Gases"-type

Table 3.1 National average life cycle water use per unit energy products (m³/TJ)

		Coal	Crude oil	Natural gas	Petroleum products	Coke	Electricity*	Heat	Gases
Water withdrawal	Direct	18.4	37.2	13.6	87.2	36.1	4306.5	609.4	7.0
	Life cycle	106.4	257.9	104.5	446.8	217.7	5263.0	901.3	42.4
Water consumption	Direct	8.9	26.3	9.5	37.9	26.2	410.9	81.2	4.5
	Life cycle	41.5	99.1	41.2	168.3	95.4	656.7	172.7	18.4
Water discharge	Direct	28.3	11.0	4.1	49.3	9.9	90.6	15.7	2.4
	Life cycle	38.4	42.8	18.7	104.3	52.7	234.2	81.8	8.6

Zhang and Anadon, 2013

*Including both coal-fired thermal power generation and noncoal power generation

energy products is the smallest. Zhang and Anadon comment that "this is because a large proportion of gases are recovered as by-products in industrial production processes, such as coking, and iron and steel making. Water use in the main production processes is not allocated to the recovered gases, since their economic values are rather small when compared with the main products, and the recovering of gases has a negligible impact on the water use of the main production process" (p. 14461).

It is also important to consider the regional differences of water use intensity, since in China there is an important geographical mismatch between the production of energy and water availability. The problem is particularly acute in what concerns the distribution of coal (by far the most dominant primary energy source) and water resources. The three provinces with the largest coal outputs are Shanxi, Shaanxi, and Inner Mongolia. These three provinces contribute more than half to the total national output of coal, and 16 per cent of thermal power generation, but are only endowed with 3 per cent of the national water resources (NBSC, 2011).

According to the World Resources Institute (WRI), as of July 2012, the Chinese government planned to build 363 coal-fired power plants across China with a combined generating capacity exceeding 557 gigawatts (for reference, installed capacity at the end of 2012 was 758 GW). Using WRI's Aqueduct Water Risk Atlas, the WRI overlaid the locations of these proposed coal plants on water stress maps for China. According to Luo, Otto, and Maddocks (2013), 60 per cent of the total proposed generating capacity is concentrated in six provinces (Shanxi, Gansu, Shaanxi, Ningxia, Hebei, and Inner Mongolia), which only account for 5 per cent of China's total water resources. On the other hand, 51 per cent of China's new coal-fired power plants would be built in areas of high or extremely high water stress. "Coal mines depend on water to extract, wash, and process the coal, while coal-burning power plants need water to create steam and cool generating systems. If all of the proposed plants are built, the coal industry – including mining, chemical production, and power generation – could withdraw as much as 10 billion cubic meters of water annually by 2015" (Luo, Otto, and Maddocks, 2013).

Zhang and Anadon (2013) also assess the life cycle water withdrawal intensity for electricity production per province (Figure 3.7), and conclude that it can be as high as 69.5 m^3/MWh in Shanghai and as low as 1.2 m^3/MWh in Qinghai. Such a large difference is driven by a multiplicity of factors, including "the mix of power generation technologies, the adoption of different cooling methods in thermal power plants, and the structure of intermediate inputs of the power sector" (p. 14461). In general, areas with poor water availability and a higher proportion of

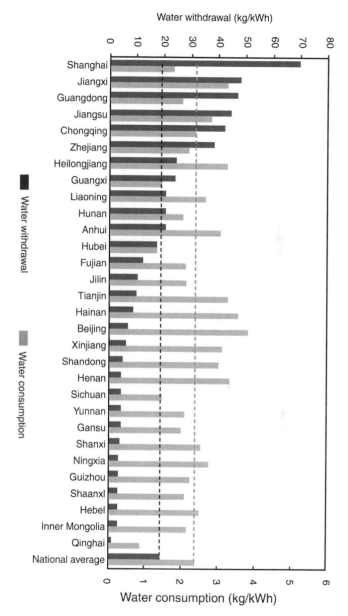

Figure 3.7 Life cycle water withdrawal and water consumption per kWh of electricity by region
Zhang and Anadon, 2013

thermal power generation tend to have lower water withdrawals and higher water consumption intensity of electricity. Places with such characteristics are mainly located in northern China where water resources are scarce, such as Beijing, Tianjin, Shandong, Henan, and Shanxi Province. In Shanghai, all electricity is produced by thermal power plants, but in Qinghai hydropower plays a dominant role. In-stream water use and evaporation from reservoirs used for hydropower generation are not considered in this study, so the water withdrawal and consumption intensity of electricity production in Qinghai are the lowest. In terms of water consumption intensity, Beijing has the highest value at 3.81 m^3/MWh (Zhang and Anadon, 2013).

Figure 3.8 shows the provincial breakdown of water withdrawal and water consumption related to energy production in China (Zhang and

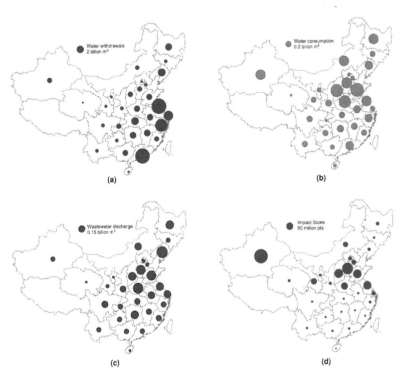

Figure 3.8 Spatial distribution of (a) life cycle water withdrawals, (b) water consumption, (c) wastewater discharge, and (d) environmental impacts related to consumptive water use of energy production in China

Note: Values are proportional to the area of corresponding circles – the scales are different.

Zhang and Anadon, 2013

Anadon, 2013). Most energy-related water withdrawals (Figure 3.8a) occur in the eastern and southern coastal regions. Water withdrawal in Guangdong, Jiangsu, Zhejiang, and Shanghai are equivalent to 23 per cent, 24 per cent, 30 per cent, and 49 per cent of their total local water withdrawal for all sectors, and account for 53 per cent of the national total energy-related water withdrawal. Electricity production and consumption is particularly high in the Yangtze River Delta on the east coast and the Pearl River Delta on the south coast, as these areas are China's major manufacturing hubs and most populated areas. The spatial distribution of energy-related water consumption (Figure 3.8b) shows considerable difference from that of water withdrawals. There is a trend of an inverse pattern between the spatial distribution of water consumption and freshwater resources. The most intense water consumption occurs in the Huang–Huai–Hai River Basin, where the most severe water shortage problems occur. The spatial distribution of wastewater discharge (Figure 3.8c) is fairly similar to that of water consumption. Figure 3.8d shows the spatial distribution of the environmental impacts caused by consumptive water use. The provinces of Xinjiang, Hebei, Shanxi, Jiangsu, Shandong, Henan, Gansu, and Inner Mongolia together bear 85.4 per cent of the total environmental impacts induced by the life cycle consumptive water use of energy production. On the other hand, in southern China, water consumption-related environmental impacts are almost negligible (Zhang and Anadon, 2013).

References

Anna, C. (2010, July 21) China oil spill doubles in size, called 'severe threat'. World Environment on NBCNEWS.com. Retrieved from http://www.nbcnews.com/id/38337393/ns/world_news-world_environment/t/china-oil-spill-doubles-size-called-severe-threat/#.VgoRuWRVhHw

Bonton, A., Rouleau, A., Bouchard, C. & Rodriguez, M. J. (2010) Assessment of groundwater quality and its variations in the capture zone of a pumping well in an agricultural area. *Agricultural Water Management*, 97(6), 824–834.

Chandel, M. K., Pratson, L. F., Jackson, R. B. (2011) The potential impacts of climate-change policy on freshwater use in thermoelectric power generation. *Energy Policy*, 39, 6234–6242.

Cheng, Q., Wang, W., Wang, H., Wang & Zhao, Z. (2012) Investigation of the heavy metal contamination of the sediments from the Yellow River Wetland Nature Reserve of Zhengzhou, China. *Iranian Journal of Public Health*, 41(3), 26–35.

Davies, E. G. R., Kyle, P., Edmonds, J. A. (2013) An integrated assessment of global and regional water demands for electricity generation to 2095. *Advances in Water Resources*, 52, 296–313.

The Economist (2010, Aug. 7) Pollution in China: raising a stink. Retrieved from http://www.economist.com/node/16744110

Gardner, E. K. (2015) Farming is driving force in drying soil in Northern China. Retrieved from http://www.purdue.edu/newsroom/releases/2015/Q3/soil-in-nor thern-china-is-drying-out-and-farming,-not-climate-change,-is-culprit.html

Gleick, P. (2009) The World's Water: 2008–2009. In Gleick, P. (Ed.) *The World's Water 2008–2009*. Washington, D. C.: Island Press.

Greenpeace (2012) Thirsty coal: a water crisis exacerbated by China's new mega coal power bases. Retrieved from www.greenpeace.org/eastasia/Global/ eastasia/publications/reports/climate-energy/2012/Greenpeace%20Thirsty% 20Coal%20Report.pdf.

Greenpeace (2014, January 23) A monstrous mess: toxic water pollution in China. Retrieved from http://www.greenpeace.org/international/en/news/fea tures/A-Monstrous-Mess-toxic-water-pollution-in-China/

Guo, J., Fang, J. & Cao, J. (2012) Characteristics of petroleum contaminants and their distribution in Lake Taihu, China. *Chemistry Central Journal*, 6(1), 92. doi: 10.1186/1752-153X-6-92

Hightower, M. & Pierce, S. A. (2008) The energy challenge. *Nature*, 452, 285–286.

Hoekstra, A. Y., Chapagain, A. K., Aldaya, M. M. & Mekonnen, M. M. (2011) *The Water Footprint Assessment Manual: Setting the Global Standard*. London: Earthscan.

Hu, D. & Xu, S. (2013) Opportunity, challenges and policy choices for China on the development of shale gas. *Energy Policy*, 60, 21–26.

Hu, F., Tan, D. & Lazareva, I. (2014, March 12) 8 Facts on China's wastewater. China Water Risk. Retrieved from http://chinawaterrisk.org/resources/ana lysis-reviews/8-facts-on-china-wastewater/

Huang, J., Xu, Q., Wang, X., Xi, B., Jia, K., Huo, S., Liu, H., Li, C. & Xu, B. (2015) Evaluation of a modified monod model for predicting algal dynamics in Lake Tai. *Water*, 7(7), 3626–3642.

IEA (2012) Key world energy statistics 2012. Retrieved from www.iea.org/p ublications/freepublications/publication/kwes.pdf.

Jun, M. (2007, June 8) Disaster in Taihu Lake. *China Dialogue*. Retrieved from https://www.chinadialogue.net/article/1082-Disaster-in-Taihu-Lake

Kahn, J. (2007, October 14) In China, a lake's champion imperils himself. *The New York Times*. Retrieved from http://www.nytimes.com/2007/10/14/ world/asia/14china.html?pagewanted=all

Kaiman, J. (2013a, February 21) Chinese environment official challenged to swim in polluted river. *The Guardian*. Retrieved from http://www.theguardian. com/environment/2013/feb/21/chinese-official-swim-polluted-river

Larson, C. (2014, Aug. 7) Something is scary in the water that irrigates many Chinese parks. *Bloomberg News*. Retrieved from http://www.bloomberg.com/ bw/articles/2014-08-07/in-china-wastewater-irrigates-parks-and-spreads-bacteria

Li, Y. P., Tang, C. Y., Yu, Z. B. & Acharya, K. (2014) Correlations between algae and water quality: Factors driving eutrophication in Lake Taihu, China. *International Journal of Environmental Science and Technology*, 11(1), 169–182.

Liang, G. & He, H. (2012) Long struggle for a cleaner Lake Tai. *China Dialogue*. Retrieved from https://www.chinadialogue.net/article/4767-Long-struggle-for-a -cleaner-Lake-Tai-

Lin, L. (2014b) Chinese countryside facing more serious drinking water crisis than cities. *China Dialogue*. Retrieved from https://www.chinadialogue.net/ blog/6960-Chinese-countryside-facing-more-serious-drinking-water-crisis-tha n-cities/en

Liu, B., Liu, H., Zhang, B. & Bi, J. (2013) Modeling nutrient release in the Tai lake basin of China: Source identification and policy implications. *Environmental Management*, 51(3), 724–737. doi: 10.1007/s00267-012-9999-y

Liu, L. (2013, February 18) Water pollution fury grows. *Global Times*. Retrieved from http://www.globaltimes.cn/content/762123.shtml

Liu, X. & Wen, Z. (2012) Best available techniques and pollution control: a case study on China's thermal power industry. *Journal of Cleaner Production*, 23(1), 113–121.

Liu, X., Ju, X., Zhang, Y., He, C., Kopsch, J. & Fusuo, Z. (2006) Nitrogen deposition in agroecosystems in the Beijing area. *Agriculture, Ecosystems and Environment*, 113(1), 370–377.

Liu, Y., Pan, Z., Zhuang, Q., Miralles, D. G., Teuling, A. J., Zhang, T., An, P., Dong, Z., Zhang, J., He, D., Wang, L., Pan, X., Bai, W., Niyogi, D. (2015) Agriculture intensifies soil moisture decline in Northern China. *Nature Scientific Reports*, 5, 11261. Retrieved from http://www.nature.com/articles/ srep11261

Luo, C. (2013c, March 23) Henan landfill threatens to pollute Beijing's drink-ing water. *South China Morning Post*. Retrieved from http://www.scmp.com/ news/china/article/1197470/mountain-rubbish-potential-threat-wa ter-resource-beijing

Luo, T., Otto, B., Maddocks, A. (2013, Aug. 26) Majority of China's proposed coal-fired power plants located in water-stressed regions. World Research Institute. Retrieved from http://www.wri.org/blog/2013/08/majority-china% E2%80%99s-proposed-coal-fired-power-plants-located-water-stressed-regions

Macau Daily (2015, August 28) Oil pollution at Zhuhai tourist spot prompts clearing efforts from officials. *Macau Daily*. Retrieved from http://macauda ilytimes.com.mo/oil-pollution-at-zhuhai-tourist-spot-prompts-clearing-efforts-from-officials.html

Meng, Y. (2012, July 9) The damaging truth about Chinese fertiliser and pesticide use. *China Dialogue*. Retrieved from https://www.chinadialogue.net/article/ show/single/en/5153-The-damaging-truth-about-Chinese-fertiliser-and-p esticide-use7

Mielke, E., Anadon, L. D. & Narayanamurti, V. (2010) Water consumption of energy resource extraction, processing, and conversion. Energy Technology Innovation Policy Research Group. Retrieved from http:// belfercenter.ksg. harvard.edu/files/ETIP-DP-2010-2015-final-4.pdf.

NBSC (2011) *China Energy Statistical Yearbook 2010*. Beijing: China Statistics Press.

NDTV (2011, Jan. 14) Acid rains make life hard in 258 Chinese cities. Retrieved from http://www.ndtv.com/world-news/acid-rains-make-life-hard-in-258-chinese-cities-444967

Philips, T. (2014, April 11) Panic after Chinese city declares tap water toxic. *The Telegraph.* Retrieved from http://www.telegraph.co.uk/news/worldnews/asia/china/10760027/Panic-after-Chinese-city-declares-tap-water-toxic.html

Piao, S., Ciais, P., Huang, Y., Shen, Z., Peng, S., Li, J. ... & Fang, J. (2010) The impacts of climate change on water resources and agriculture in China. *Nature*, 467(7311), 43–51.

Radio Free Asia (2014a) Authorities investigate pollution in China's iconic Erhai Lake. Retrieved from http://www.rfa.org/english/news/china/erhai-03262014145618.html

Ren, T., Zhao, L., Sun, B. & Zhong, R. (2013) Determination of lead, cadmium, copper, and nickel in the Tonghui river of Beijing, China, by cloud point extraction-high resolution continuum source graphite furnace atomic absorption spectrometry. *Journal of Environmental Quality* 42(6), 1752–1762.

Rothman, A. (2006) Thirsty China: Its key resource constraint is water. CLSA. Retrieved from http://chinawaterrisk.org/wp-content/uploads/2011/04/Thirsty-China.pdf

Sha, L. (2014) Blasts blamed for toxic water contamination. *Global Times.* Retrieved from http://www.globaltimes.cn/content/854394.shtml

Sun, R., Chen, L., Chen, W. & Ji, Y. (2013) Effect of land-use patterns on total nitrogen concentration in the upstream regions of the Haihe river basin, China. *Environmental Management*, 51(1), 45–58. doi: 10.1007/s00267-011-9764-7

Thorburn, P. J., Biggs, J. S., Weier, K. L. & Keating, B. A. (2003) Nitrate in groundwaters of intensive agricultural areas in coastal northeastern Australia. *Agriculture, Ecosystems and Environment*, 94(1), 49–58.

Tilt, B. (2013) The politics of industrial pollution in rural China. *The Journal of Peasant Studies*, 40(6), 1147–1164. Retrieved from http://www.tandfonline.com/doi/pdf/10.1080/03066150.2013.860134

Vidic, R. D., Brantley, S. L., Vandenbossche, J. M., Yoxtheimer, D. & Abad, J. D. (2013) Impact of shale gas development on regional water quality. *Science*, 340(6134), 1235009.

Wang, S., Ma, X., Fan, Z., Zhang, W. & Qian, X. (2014) Impact of nutrient losses from agricultural lands on nutrient stocks in Dianshan Lake in Shanghai, China. *Water Science and Engineering*, 7(4), 373–383. Retrieved from http://www.sciencedirect.com/science/article/pii/S1674237015302957

Wei, F. (2012, October 4) A poisoning exposed. *China Dialogue.* Retrieved from https://www.chinadialogue.net/article/show/single/en/4864-A-poisoning-exposed

Wong, E. (2013, March 2) Spill in China underlines environmental concerns. *The New York Times.* Retrieved from http://www.nytimes.com/2013/03/03/world/asia/spill-in-china-lays-bare-environmental-concerns.html?_r=0

Wu, W., Sun, R. & Peng, Y. (2013) Chemical leak into river puts focus on plant. *China Daily.* Retrieved from http://usa.chinadaily.com.cn/china/2013-01/11/content_16104337.htm

Xiao, R., Bai, J., Huang, L., Zhang, H., Cui, B. & Liu, X. (2013) Distribution and pollution, toxicity and risk assessment of heavy metals in sediments from urban and rural rivers of the Pearl River delta in southern China. *Ecotoxicology*, 22(10), 1564–1575. doi: 10.1007/s10646–10013–1142–1141

Xinhuanet (2015, July 27) China sees less acid rain in the first of 2015. Retrieved from http://news.xinhuanet.com/english/2015-07/27/c_134452139.htm

Yin, P. (2015, May 21) Solving the water problem: China lays out a blueprint to curb water pollution. *Beijing Review*. Retrieved from http://www.bjreview.com.cn/print/txt/2015-05/18/content_688338.htm

Zhang, C. & Anadon, L. D. (2013) Life cycle water use of energy production and its environmental impacts in China. *Environmental Science and Technology*, 47(24), 14459–14467. doi: 10.1021/es402556x

Zhang, Y., Ma, R., Duan, H., Loiselle, S. & Xu, J. (2014) A spectral decomposition algorithm for estimating chlorophyll-a concentrations in Lake Taihu, China. *Remote Sensing*, 6(6), 5090–5106.

Zhang, T., Xie, X., & Huang, Z. (2014) Life cycle water footprints of nonfood biomass fuels in China. *Environmental Science and Technology*, 48(7), 4137–4144.

Zhao, X., Chen, L. & Zhang, H. (2013) Nitrate and ammonia contaminations in drinking water and the affecting factors in Hailun, northeast China. *Journal of Environmental Health*, 75(7), 28–34.

4 The consequences of water pollution

Antibiotics waste

China is the largest producer and user of antibiotics in the world (Zhu, 2013). Zhang et al. (2015) reported that 162,000 tons of antibiotics were consumed in China in 2013: 84,240 tons of antibiotics were given to animals to prevent disease, and 77,760 tons to people (Zhang 2015, Table 2). This is about nine times more than in the U.S., where 14,600 tons of antibiotics are consumed by animals, and 3,290 tons are consumed by people (Zhang et al., 2015, Table 2). Almost all antibiotics end up in the water bodies after excretion by animals and humans. Radio Free Asia (2015) reported that according to Ying Guang-Guo, a professor in environmental chemistry and ecotoxicology, "the same antibiotics that end up in rivers and fields through wastewater then return to human bodies through aquatic and agricultural products, forming a vicious circle". According to some estimates, in China there are "one million deaths a year from antibiotic-resistant infections, which occur when the microbes adapt to the antibiotics when there is not enough dosage to kill them" (Radio Free Asia, 2015). Some of the bodies of water with the highest amounts of antibiotics are the Dongting Lake in central China where researchers found 3,440 tons of substances in 2013, and the Yellow, Huaihe, and Yangtze downstream river basins where researchers found 3,000 tons of antibiotics in 2013 (Figure 4.1).

An interesting side story to this is that one of the U. S. doctors interviewed in Radio Free Asia (2015), Wan Yanhai, fled China after he "blew the whistle on the spread of the HIV/AIDS epidemic through rural blood-selling schemes in poverty-stricken Henan Province". He explains that one of the reasons for the over-prescription of antibiotics occurs because many of the patients in China have a set of expectations about prescriptions and insist on having a prescription, no matter what the ailment, whenever they see a doctor; "otherwise they don't see the

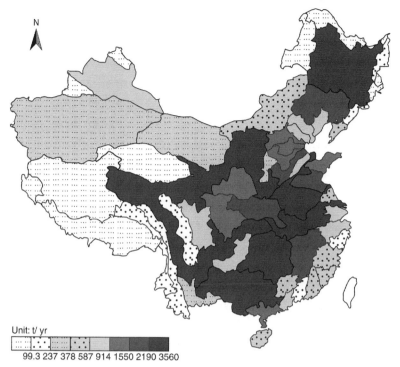

Figure 4.1 Antibiotic emission map of China
Zhang et al., 2015

point [...] if the patient asks for antibiotics, they will prescribe them". Wan also states that many doctors satisfy those demands because every prescription boosts their incomes. He asserts that the government is fully aware of the link between doctors' incomes and pharmaceutical companies, as well as the numerous antibiotics used in poultry farms, but the government's efforts to control this have been minimal.

Radio Free Asia (2015) also cites a study by Shanghai's Fudan University that tested 1,064 children aged 8–11 in Shanghai, Jiangsu, and Zhejiang provinces for 18 antibiotics, and discovered them present in the urine samples of 60 per cent of the children. In 2014, the WHO identified antibiotic resistance as a major threat to public health, leading to longer periods of sickness and higher mortality rates. Meanwhile, Hui (2015) quoted Zhou Xiaoqing as saying that "300,000 children in China have become deaf through overuse of antibiotics". However, despite widespread evidence of environmental contamination, antibiotics were not included in the tests of drinking water quality (Radio Free Asia, 2015).

Health impact of industrial and domestic wastewater

Carlton et al. (2012) found that China's rapid economic growth over the past decades brought a shift in health priorities, culminating in infectious diseases usually associated with poverty being gradually displaced by chronic illnesses. Yet, the researchers also reported that the traditional causes of illnesses, including infections from unsafe water and poor sanitation and hygiene continue to exist. Unsafe water and poor sanitation and hygiene can cause illness through various means: "Drinking water can be contaminated with biological or chemical agents, soil, water or fomites can be contaminated with faeces, and, if water resources are poorly managed, they can become vector habitats" (p. 579). However, it is also important to recognize that such contamination is unevenly distributed across the country's diverse landscapes as a result of regional differences in urbanization, economic development, and environmental factors. The authors write: "Countrywide measures of important infectious diseases conceal important regional and socio-economic disparities that, although widely recognized in China, have been poorly documented, particularly for diseases resulting from environmental pollution" (p. 578). Consequently, the authors report that although the water and sanitation infrastructure has improved dramatically over the past decades in China, "access to safe water and good sanitation varies markedly throughout the country, which suggests that some population groups bear greater risks of disease than others" (p. 578). The authors note that in 2008 unsafe water and poor sanitation and hygiene accounted for 62,800 deaths and 2.81 million disability-adjusted life years (DALYs) in China. Rural residents, who represent 60 per cent of China's population, are particularly vulnerable (Figure 4.2).

Carlton et al. (2012) conclude that the "distributions of total and disease-specific DALYs attributable to unsafe water and poor sanitation and hygiene showed substantial geographical and socioeconomic disparities. Unsafe water and poor sanitation were found to be particularly detrimental to the health of young children and accounted for 61,200 deaths and 2.33 million DALYs in children under five, predominantly attributable to diarrhoeal diseases" (p. 581). Geographically, the disease was concentrated in China's poorest inland provinces. The researchers stress that their findings indicate "the need for further work to increase access to improved water and sanitation and to reduce disparities in the

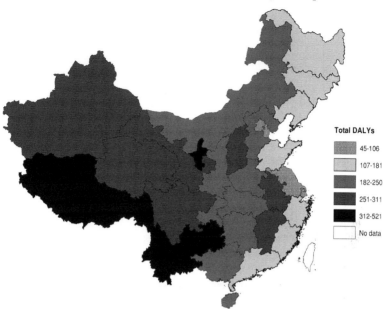

Figure 4.2 The distribution of disability-adjusted life years (DALYs) attributable to unsafe water and poor sanitation and hygiene, by province,
in 2008
Note: Vector-borne infections include dengue, malaria and Japanese encepha-
litis. Helminthiases include ascariasis, hookworm infection, trichuriasis, and
schistosomiasis.
Carlton et al., 2012

disease burden attributable to poor sanitation and unsafe water supplies" (p. 584).

Li (2015) concurs with many of these findings, writing that the massive shifts of millions of people from rural to urban areas and con-
sumerist lifestyles result in 300 million tons of waste each year, placing
a severe stress on underdeveloped public waste management services.
Usually, the urban waste management services collect unsorted municipal solid waste (MSW) that is disposed of in landfills or waste incinerators around the periphery of the city or deep into the countryside.
According to Li, government waste services do not have the capacity
to operate a recycling system, so even if the waste is separated at the
source, in the end it is dumped into one load and sent to landfills and
waste incinerators. Organic matter in these landfills cannot decompose
properly, which results in the release of methane, a potent greenhouse

gas. Besides, the urban waste stream is an inefficient fuel for incineration. Furthermore, Li writes:

> In addition to the poor waste collection infrastructure, investment, and enforcement, the current waste system in China perpetuates social inequalities for rural-to-urban migrants who enter urban spaces with low socioeconomic statuses. Landfills and incubators are pushed to the outskirts of the city where poor migrants live, bringing along toxic fumes of incineration, disturbances from trucking of waste, and pollution of water, air, and soil. This leaves the wealthier inner city areas relatively clean, while the pollution impacts of their waste are exported to small towns and poor communities that are socially, politically, and economically marginalized from the city.

Li also divulges that a significant number of migrants dominate urban recycling of any material perceived of value, which results in them handpicking through rubbish bins, and moving along streets collecting items such as paper, cardboard, plastic, metals, electronic waste, or anything that is recyclable. It is estimated that about 3.3–5.6 million people in cities throughout China are involved in this informal recycling sector and are responsible for recycling about 17–38 per cent by weight of municipal solid waste. According to Li, activists say that in Beijing "there are around 200,000 informal collectors working seven days a week, collecting around 30 per cent by weight of the total MSW". These informal collectors who take advantage of the local governments' inability to provide adequate infrastructure, services, and education for a more formal recycling system earn very little for their efforts, are older, and often live in very poor conditions.

Cancer villages

Delang (2016) discussed the impact of air pollution on what have become known as cancer villages, but did not discuss the impact that water pollution plays in creating this unfortunate phenomenon. According to Liu (2010) in 1998, China Central Television (CCTV) and *Sheghuo Shibao* (*Life Times*) were among the earliest Chinese media to reveal the existence of cancer villages. Their reports focused on the pollution on the Hai River flowing through Tianjin and neighbouring Hebei, where chemical oxygen demand (COD) was over 1,300 mg. Liu noted that 25 mg are sufficient to downgrade the water to Grade V, the most polluted level in the five-grade Chinese water quality classification

(Chapter 2). Liu asserts that water contamination from industrial pollution is the primary cause of cancer villages, and finds that their location closely correlates with China's major rivers.

> Cancer villages tend to cluster along the major rivers and their branches. These rivers have supported high population density for thousands of years. They are also the prime location choices for industries that require cheap water, labour, and transportation. Many industrial parks have found homes along these rivers, which are now heavily polluted. While these industries have contributed to rapid GDP growth in their regions, this growth has been achieved at the expense of the health and lives of poor villagers, in many of the village economies. The largest concentrations of cancer villages are located along the lower reaches of the Yellow River and the Changjiang River and Pearl River deltas, the two most developed areas of China. Inland concentrations of cancer villages are found along the Yellow, Huai, and Changjiang rivers, as well as the Beijing-Hangzhou Grand Canal. Sewage monitoring inspections by the Ministry of Water Resources found all provinces in the Huai river basin guilty of water pollution.
>
> Liu, 2010, p.8

Liu (2010) reports that in Guangdong Province, China's third most developed province behind Beijing and Shanghai, cancer villages exist in all four regions (the Pearl River Delta, East, West, and North Mountainous regions). Liu demarcates Wongyuan County in the Mountainous region as being one of the worst cases of cancer villages in China. The iron and copper sulphide strip mining in the region since 1970 has resulted in "serious soil erosion and landslides that have dumped cancer-causing cadmium, lead, and other heavy metals into the water system and soil down the mountain". Eventually, the lush riverside agricultural settlements in the valleys became cancer villages. The village of Liangqiao, closest to the mines, is believed to be the worst polluted. However, the death toll has been the heaviest in the most populated village of Shangba farther down the river, where over 250 villagers around the age of 50 died of cancer between 1978 and 2005 (Liu, 2010).

Liu identified several other regions in the Chinese countryside with high cancer rates (Figure 4.3). In particular, Shenqiu County (in Henan Province) has the largest clusters of cancer villages. Numerous media outlets have focused on water pollution as the reason for the increased cancer rates in the county. The village of Huangmengying, in

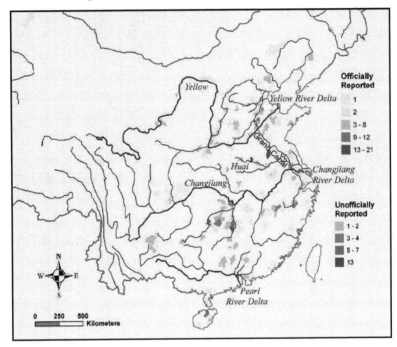

Figure 4.3 Major rivers and counties with cancer villages in China, 2009
Liu, 2010

Shenqiu County does not fare much better. "The rate of death in
Huangmengying Village increased from 5 in 1,000 in 1990, to 8 in 1,000
in 2004. Liver, rectum and stomach cancers – mostly cancers of the
digestive system – claimed the lives of 118 villagers, about half of all
deaths, out of 2,400 residents between 1994 and 2004. The youngest, a
one-year-old, died of intestinal cancer" (Liu, 2010). The local environmental protection agency believes the problems are a result of the water
contamination spewing from factories in industrial cities upstream
(Wang, 2004).

Another case reported by Liu (2010) is that of Xingang Village
(Yandu District, Jiangsu Province). It used to be called Taohua Yuan
(land of peach blossoms) with its many rivers, productive rice fields,
and lush vegetation, and at one time it was considered one of the
prime getaways and tourist spots in the region. In 2001, things changed
following the industrial park that was built on the village's farmland.
Toxic gas and water from the chemical factory poisoned the land. Rice
yields severely dropped because of the toxicity, and the pigs died in

large numbers, making the villagers quit pig farming. Subsequently, the village fell into poverty and in a few years 55 villagers developed cancer; 40 of them died from it.

The photographer Souvid Datta recently documented water pollution in different provinces. Among Datta's photographs was that of a factory (Figure 4.4) with the caption "The Shuogang group steel factory, Beijing. Despite the government promising to close all major polluting factories within city limits following the 2008 Olympics, several are still operating behind closed doors. At dawn every day the factory wastewater pipe illegally spews hazardous chemicals into a local dried up lake. The accumulated green and brown deposits contain poisonous metal deposits, which are visible here" (Datta, 2015).

In Yinzhou (Zhejiang Province), Datta followed up with the Youngor textiles factory, the subject of an investigation by Greenpeace in 2011. Associated with global brands such as Adidas, the company had been exposed for dumping wastewater into the local Fenghua River – which provides for the village of Rongjianqui. Three years after promising to improve its practices and cease illegal dumping, Datta noticed the wastewater pipe had been moved across the river into a difficult to access location only a few kilometres upstream. Subsequently, the dumping hours had been switched from noon to dawn. He was told

Figure 4.4 The brutal reality of life in China's most polluted cities
Datta, 2015

that since 2011, 12 people had died of cancer in Rongjianqui (Eiferman, 2014).

Cao et al. (2015) looked at the relationship between water quality and oesophageal cancer (EC) in the rural area in Anyang (Henan Province), using a dataset of 3,806 individuals utilising 550 drinking water sources in 92 townships. The cases included 531 EC patients, and the controls included 3,275 people aged over 90 years and free from EC. Since in China population migration is strictly controlled by government regulations, the location of residents at the time of observation will likely reflect their exposure to drinking water pollution throughout their lifetime. Because of this, researchers can obtain reliable data about the relationship "between water quality and oesophageal cancer (EC) in the rural areas where most of the inhabitants still rely on springs, cisterns, rivers, wells or tap water as their drinking sources" (p. 2). Their results show that pollutants in drinking water were associated with high risk of EC. "The geographic, ecologic and economic factors might elevate the pollutant concentration in drinking water. In poor water resource areas, drinking water stored in cisterns might accumulate nitrosation precursors. Pollutants from the coal industry were intercepted by rivers and seep into groundwater to contaminate drinking water resources. Therefore, in remote villages without tap water provision and areas at river flow path turns, the EC incidence is higher" than that in other regions (p. 8).

Impact of contaminated water on health

In 2013, an official apology was issued and an investigation was launched after the claims by Dongchuan environmental bureau chief, Wang Jieyn, that the water in a white polluted stream was not only safe for irrigation, but also safe to drink (Luo, 2013d). Wang's statement on the quality of the water was met with disbelief after photos had emerged online showing a villager in the Dongchuan district of Kunming carrying buckets of water to be used for drinking from the stream later dubbed "milk river" due to its white colour. A district agricultural official derided Wang's conclusion stating "the environmental bureau is both law enforcer and interested party". The local media reported that an environmental expert found the heavy metal contents of the stream were approximately "50 times higher than claimed by the environmental bureau". The Dongchuan district head, Lu Ping, issued an apology and an acknowledgement that the government was at fault. Ping admitted that the water pollution was due to wastewater discharge from mineral processing factories and promised to crack down on the factories responsible for the damage.

More media attention was given to the plight of China's water pollution after a veteran policeman had rescued a 14-year-old girl out of the sewage and bile from a river in Wenzhou City (Zhejiang Province). Kao (2013b) reports that later in the day the police officer, Zhang Guangcong, began coughing, vomiting and suffering from burning eyes, skin irritations, and dizziness. He was admitted the following day to the Third People's Hospital where doctors diagnosed him as having a severe infection of the lungs; more than likely "contracted from the bacteria-infested river". Although Zhang was visited by the county's chiefs of police and education, the Head of the Environmental Protection Bureau did not pay him a visit, much to the chagrin of angry citizens who subsequently criticised the country's growing water pollution online and called for the Bureau to issue a "formal apology" to Zhang.

According to Kaiman (2013a), an eyeglass-retailer executive, Jin Zhengmin, from Rui'an City (Zhejiang Province), symbolised the public's anger when offering the city's environmental protection chief, Bao Zhenming, more than CNY 200,000 to swim for 20 minutes in a highly polluted local river. Two days later another offer was made with a reward of CNY 300,000 for a 30-minute swim (Luo, 2013). Zhengmin blamed the local rubber shoe factory for the river's deteriorating standard. He posted the dare to his microblog beneath the pictures he collected showing the waterway overflowing with discarded aluminium cans, polystyrene boxes, and paper lanterns. Bao declined the offer, and Rui'an's city government claimed the river's pollution was caused by individuals, not factories, and should be attributed to overpopulation.

A Pailian villager in Zhejiang, Chen Yuqian, 60, also offered an environmental official money to see if he had the guts to swim in a polluted river, but Chen was attacked by 40 angry villagers afterwards (Luo, 2013b). Chen Yuqian's daughter posted the photos online showing her father's scratched and bruised face. She also commented that the assault on her father was orchestrated by village officials in retaliation for previous petitions about water pollution. In a petition letter also posted online, Chen accused Pailian officials of teaming up with two local paper mills to illegally dump industrial waste into waterways. The petition claimed that the dumping had polluted the drinking water of more than 200 villagers. The officials denied culpability and claimed that the paper mills were operating in accordance with the regulations.

Chinese citizens flooded the internet with enraged comments and photos of the massive environmental problems depicted by pictures of polluted streams and rivers, and dead fish and animals floating downstream (Garland, 2013). Nothing was more staggering than the view of

16,000 bloated pig carcasses floating downstream and being fished out of the Huangpu River, near Shanghai, and its tributaries. Davison (2013) interviewed a 48-year-old fisherwoman who recalled the time when she was a child and could splash about in the river on hot summer days. Her memories had since been quashed as she stared into an inky black river, covered in a slick of lime green algae, and smelling like a blocked drain. The fisherwoman said, "Look at the water, who would dare to jump in?" She also half joked that there are more pigs than fish in Jiapingtang River, as a dead piglet bounced against the shore. This area of Zhejiang Province, 60 miles from Shanghai, became a place of intense media scrutiny after the pigs had been found in the tributaries of the city's river. In the 1980s, the pig industry blossomed in Jiaxing, and pig farming continued to prosper for many decades in this region. According to Davison, in 2012, China produced and consumed half the world's pork. However, "with a mortality rate of 2–4 per cent, up to 300,000 carcasses need to be disposed of each year" (Davison, 2013). With the growing number of pig farms, lack of funding and land to build more plants, some believed that farmers were occasionally throwing dead pigs into the rivers. The sight of dead swine floating among plastic bottles and flotsam accompanied by the stench of putrefying pig flesh is bad enough, but as reported by *The Economist* (2013) it was even more alarming for Shanghai's residents that the same waterway supplied between 20–30 per cent of the city's tap water.

A state-run exposé by China Central Television (CCTV) showed how illegally processed pigs had made their way into the market for years. Even though farmers were required legally to send animals that die of disease or natural causes to processing pits, black market dealers were known to intercept the chain, butcher the hogs and sell the meat. Davison (2013) writes that a Jiaxing court sentenced three such butchers to life in prison; the offenders had processed 77,000 carcasses. Apparently, the crackdown caused black market traders to stop buying the dead stock, and farmers resorted to dumping the dead animals. A person in custody for suspicion of dealing in dead pigs told CCTV "there was a 100 per cent correlation between his arrest and the dead pigs incident". Tests showed that the pigs carried porcine circovirus, a common disease among pigs. The disease is not known to infect humans, and the Shanghai water supply authorities tried to quell the public outcry and insisted that drinking the water was safe, although few believed it. Nothing made it more evident to Chinese citizens who crammed the internet with their concerns that there was a serious water pollution issue than seeing thousands of dead pigs, unnaturally discoloured water, and the fetid smell that accompanied it all. This discontent has not been lost on China's top leaders who have pledged to more aggressively tackle China's pollution challenges in coming years (Garland, 2013).

Food safety and pesticides

Huang (2013) writes: "The widespread production and consumption of toxic chemicals in industrialisation and agricultural production have polluted water and air, and contaminated farmland, contributing to the emergence of as many as 400 so-called cancer villages". Huang also reiterates that the long-term health and environmental consequences of China's rapid industrialisation, combined with the inadequate regulations, is one of the explanations why one out of four people in the most populous nation in the world lacks access to safe drinking water. The WHO estimated in 2013 that nearly 100,000 people die annually from water pollution-related illnesses in China. Apart from the fact that 70 per cent of China's lakes and rivers are polluted, and almost 40 per cent of those rivers are deemed seriously polluted, a 2011 study estimated that annually more than 94 million people in China become ill because of bacterial foodborne diseases, and about 8,500 people die from them. Huang posits this number is likely an underestimation of the country's food safety crisis because the statistics on health conditions caused by tainted food are often excluded. A 2011 study by Nanjing Agricultural University concludes that 10 per cent of the rice sold in China contained excessive amounts of cadmium, and researchers estimate that as much as 70 per cent of China's farmland is contaminated with toxic chemicals (Huang, 2013).

Ivanova (2013) notes that the growth of the Chinese middle and upper classes has resulted in the increased consumption of beef and pork, and the water requirements of cattle and pigs are extremely high. Food safety has become a major concern among Chinese citizens who are worried by such high profile food safety scandals as deadly melamine milk (Huang, 2014), recycled gutter oil (Song, 2013), fake beef (Kaiman, 2013b), exploding watermelons (Watts, 2011), and more; which created "public ridicule and ire in a political system that has vowed to serve the people" (Ivanova, 2013). Authorities tried to establish national standards for water and soil, and to improve the quality of the food, but for the time being the low quality of the water available to farmers constrained their ability to grow healthy food.

Ivanova interviewed farmers in Shandong, a prime food-growing province in the lower reaches of the Yellow River. One of the farmers pulled a hose into a water pump powered by a truck with a belt-drive.

Once the engines started, the makeshift machine filled the hose with turbid water from the nearby canal where a pharmaceutical factory dumped its rancid effluent moments earlier. The farmer explained to Ivanova: "There's no water source except for this dirty water. We have to use it". He informed her that even when the water turned black the prior month and most of the crop died after being irrigated with it, the remains that did not wither were sent to the market. Thereafter, the author writes:

> The farmer's plight underlies a dirty truth about China's fast development: the nation's rivers, lakes, and falling water tables are enduring deficits of clean water that often force farmers to grow food using water that is tainted with heavy metals, organic pollutants, and nitrogen. Much of China's water is so contaminated that it should not even be touched, yet tremendous amounts of the grains, vegetables, and fruits that are served in homes and restaurants, as well as textiles that are sold in markets are irrigated with untreated industrial wastewater.

Hu Kanping of the non-profit Chinese Ecological Civilization Research and Promotion Association in Beijing says ironically, "I have seen farmers in Hebei use contaminated water because there's nothing else to use. Farmers won't eat what they produce. They have fields for themselves and fields for the market" (Ivanova, 2013). At the same time, as mentioned above, a significant portion of China's water pollution comes from agriculture – fertilisers, pesticides, and livestock waste that are dumped into rivers, lakes, and reach underground aquifers. Consuming 31.4 per cent of the total global amount of fertilisers, China has become the largest producer and consumer of fertilisers and pesticides in the world.

Agricultural runoff is called nonpoint source pollution (NPS) because it can't be traced to only one source, unlike pollution from an industrial site or factory. Ivanova stresses that "agricultural nonpoint source pollution is the dominant source of water pollution in China, and it also serves to increase soil erosion and reduce the productivity of the land". Ivanova reports that according to Wang Dong, a senior expert from the Chinese Academy for Environmental Planning, animals produce about 90 per cent of the organic pollutants and half of the nitrogen in China's water. We should add that NPS is the most difficult and costly to tackle and clean up.

The impact of soil and water pollution on food safety nationwide has been studied by Lu, Song et al., (2014). Lu, Song, et al. confirm

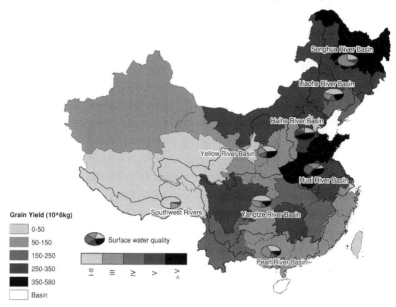

Figure 4.5 Distribution of water quality and grain yield, 2010
Lu, Song, et al., 2014

previous findings that cancer villages tend to cluster in Eastern China's
grain producing region, and call for a boost to more stringent food
safety policies (Figure 4.5). The authors discuss how soil and water
pollution have historically impacted food safety and how they impose a
major threat to human health and well-being. They write:

> Water scarcity, pesticide over-application, and chemical pollutants are
> considered to be the most important factors impacting food safety in
> China. Inadequate quantity and quality of surface water resources
> in China have led to the long-term use of waste-water irrigation to
> fulfil the water requirements for agricultural production. In some
> regions this has caused serious agricultural land and food pollution,
> especially from heavy metals. It is important, therefore, that issues
> threatening food safety, such as combined pesticide residues and heavy
> metal pollution are addressed to reduce risks to human health.
>
> p. 5

Lu, Song, et al. acknowledge that fertilisers and pesticides have been
playing a vital role in the success of modern food production for decades, and that studies have proven that fertiliser and pesticide use

contribute greatly to improved grain production. However, inefficient use of pesticides and inadequate management of pesticide application in food production constitute potential occupational hazards for farmers and environmental risks for agricultural ecosystems. The authors cite WHO (1990), which reported that there were millions of cases worldwide of unintentional occupational poisoning by pesticides, and provided evidence that pesticides are to be held responsible for severely affecting human health. Yet, among all pollutants, the researchers identified heavy metals as the greatest risk to food safety in China. They write:

> The main sources of heavy metals in farmland soils include mining and smelting, sewage irrigation, sludge reuse and fertiliser application. Due to extensive and nonstandard production processes of some mining and smelting enterprises, large quantities of heavy metals affect farmland through wastewater irrigation, waste transportation, sludge application and atmospheric deposition which has been shown to be particularly important in Southern China with abundant mineral resources.
>
> p. 6

Lu, Song, et al. conclude that the current state of China's integrated food safety policies are poor in regards to soil and water pollution. The researchers stress that there is a need for an integrated nationwide survey and information infrastructure in order to thoroughly investigate soil and water pollution, and its impacts on food safety. The authors suggest that a nationwide monitoring network be established to "collect real-time information on emission sources, major pollutants, and their distribution pathways in different environmental media" (Lu, Song, et al., 2014, p.11). In addition, "a food safety monitoring system should be established, along with the development of portable detection devices that work more quickly and efficiently to provide more comprehensive information of food safety" (p. 11). They also advise that:

> a food supply chain tracking system from field to fork should be established to make the information of food production, processing, transportation and storage open and transparent, and to take precautionary approaches to avoid the spread of contaminated food. Improving monitoring, regulatory oversight and more government transparency are needed to better estimate the potential risks of contaminated water, soil and poor sanitation and hygiene on human health.
>
> p. 11

One solution offered by Lu, Song, et al. is to set up a portfolio of policy actions. The researchers contend that there is a vital need to strengthen the coordination and cooperation between the different sectors, and to improve inter-agency collaboration owing to the large number of departments with roles in managing water, soil, food and health. Given that there are seven river basin conservancy commissions in China, the researchers advise that a new environmental protection department be created under the river basin conservancy commissions "to integrate water pollution, water resources transfer and soil erosion management" (p. 12). At the farm level, Lu, Song, et al. also suggest that farmers apply fewer agrochemicals, while the government should foster the development of clean technologies, and promote alternatives to the highly toxic chemicals through such things as tax rebates. Finally, high-risk industries should be relocated far away from food production regions.

In 2013, the authorities in Guangdong tested 18 batches of rice for cadmium levels during quarterly spot-checks, and found that eight contained excessive amounts of the carcinogenic heavy metal. Samples of the tainted rice were taken in the provincial capital, Guangzhou, and revealed readings of between 0.21 and 0.4 mg of cadmium per kg, in excess of the national limit of 0.2 mg (Duan, 2013). Studies revealed that as early as 2011, excessive cadmium levels had been found in around 10 per cent of the rice sold throughout China. Mounting public pressures for transparency over environmental scandals resulted in authorities taking such rare actions as naming the rice producers whose products contained excessive amounts of cadmium (Radio Free Asia, 2013b). Environmental activist Yang Yong posited that the rice most likely had been contaminated by the water used to irrigate the rice paddies. Yang explains, "Heavy metals can be found in water and in soil, and can be transferred into food. This can have a huge impact as it accumulates in the human body" (Radio Free Asia, 2013b). Xue Shikiu, a water resources management expert at the University of Florida concurs with this viewpoint and adds that there are three main sources of the heavy metal contamination of crops:

1   natural minerals which permeate into the water supply through weathering;
2   industrial pollution;
3   pollution derived from various sources during agricultural production.

Experts suspected that the farmers' fertilisation of the fields could be a problem. This was confirmed through a 2013 investigation by Radio

Free Asia's Cantonese Service in Guangdong, which showed that local farmers used trash from nearby landfill sites as compost, mixed it with commercial fertilisers, and spread it across their fields (Radio Free Asia, 2013b). The problem is not limited to Guangdong Province, and to using trash as compost. Rice grown near Dongting Lake in Hunan Province was also found to have excessive levels of heavy metals. Activist Yong believes the contamination in Hunan could be linked to China's biggest centre for phosphate mining, which is located on the border with nearby Guizhou. "There are large amounts of heavy metals in the wastewater and slurry produced by the phosphate mining industry" (Radio Free Asia, 2013b).

Activism

According to a 2005 Gallup Poll, only 28 per cent of Chinese people think that water pollution is not a problem where they live (Figure 4.6) (Burkholder, 2005). However, these data are skewed by the fact that most Chinese still live in rural China. In urban areas, only 12 per cent think that water pollution is not a problem, and in the ten largest cities this total drops to 8 per cent. In the top ten cities, 34 per cent believe that water pollution is either somewhat serious or very serious, four times more than those who say it is not a problem. Clearly, there

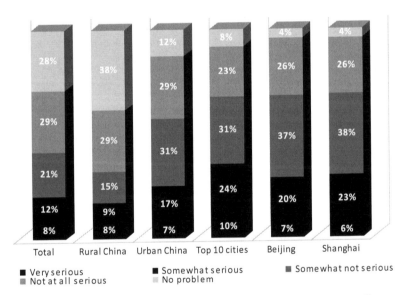

Figure 4.6 Concern over water pollution according to a 2005 Gallup poll
Burkholder, 2005

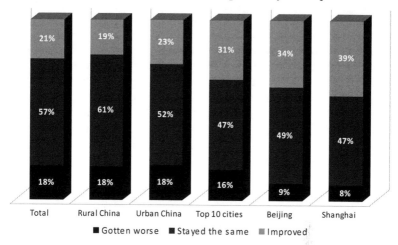

Figure 4.7 Change in the perception of water pollution according to a 2005
Gallup Poll
Burkholder, 2005

is considerable concern about the quality of the water, particularly in
urban areas. On the other hand, the same Gallup poll showed that
most people felt the pollution problem was improving, in particular in the
top ten cities where the problem was felt as more serious (Figure 4.7).

Concerns over water pollution have resulted in a number of protests,
which in turn have prompted government action. One such example
comes from the eastern city of Nanjing. Due to the complaints from
local residents, authorities in Nanjing investigated reports that factories
along the Yangtze River continuously dumped toxic effluent into the
water (Radio Free Asia, 2013a). The investigation confirmed that seven
factories near Lianmeng village in Nanjing's Qixia district were
disposing large quantities of pollutants into the river. According to the
Yangtze Evening News, "To test the level of pollution, [villagers]
bought two large, healthy Crucian carps and put them into the tank
full of the polluted water. [...] In less than 10 minutes, both fish were
floating dead on top of the water". Environmental activist Liu
Guanghua stated, "It's not just the chemical factories emptying waste
there. Our tests have shown in the past few years that toxic effluent
sometimes pours into the Yangtze from the Qinhuai River sluice gates.
As soon as they open the sluice gate from the Qinhuai River, all the
fish in the Yangtze River die off. Downstream of this, there are two
intake pipes for a treatment plant for Nanjing's tap water" (Radio Free
Asia, 2013a).

Unfortunately, activists such as Guanghua, who confronted authorities and vested commercial interest over pollution, have frequently been subjected to revenge attacks and government harassment. For example, authorities in Jiangsu Province arrested and sentenced environmentalist Ji Shulong to two years imprisonment for "obstructing official duty, stirring up trouble, and picking quarrels". Nevertheless, Radio Free Asia (2013a) reported that officials have reluctantly admitted that the country is facing a grave environmental crisis with more than half of its cities affected by acid rain and one-sixth of its major rivers too polluted to even water the crops.

Liu (2010) imparts that many of the farmers began to feel hopeless after the many failed attempts to fight the polluters, and most gave up. The educated, able-bodied villagers, particularly the young, have a tendency to move away, leaving behind those who are less capable of battling authorities. When they do fight against the polluters, desperate villagers sometimes gather and block traffic to the factories or tamper with their water supply systems to garner attention. This sets the conditions for riots to occur when the police try to arrest the protest leaders. Village leaders who organise local protests may be subjected to being shut down by authorities, receiving harsh penalties, or being violently assaulted. For example, Wang Linsheng in Shenqui (Henan Province), was Huangmengying Village's CCP secretary and made the mistake of informing the media on how many villagers had died of cancer. Not only was Linsheng fired, but he was also accused of the serious crime of leaking state secrets (Liu, 2010).

Although rare, there are some success stories as well. Here is one example:

> Luo Liquan was a millionaire fish-farmer before pollution killed his fish. He and Wang Xiufeng, co-founder and director of the Chongqing Green Volunteers Alliance, have been leading the NGO Chongqing Green Volunteers Alliance to help cancer villagers and gather evidence to fight the polluters. The villagers were so helpless that when Wu Dengming, one of the volunteers, came to investigate the pollution problem, over 100 villagers kneeled down at the village entrance to greet him, even though some local officials pressured villagers not to leak any information to the volunteers. Though harassed, assaulted, and jailed for "disturbing social life," the volunteers were able to collect credible evidence for legal actions against the polluters. After repeated media exposure, the Chongqing Environmental Protection Bureau ordered chemical factories in western Chongqing to suspend their operations for

inspection, and only three of them were allowed to resume production.

Liu, 2010

Deteriorating air and water quality, the impact of heavy metals from mining and industry on the environment, and numerous public health scandals have prompted many Chinese to become more involved in environmental protection. People are calling for more transparency over environmental standards, including water quality. In 2012, China issued new regulations setting new safety standards for public drinking water suppliers (see Chapter 5), boosting the number of items to be tested to 106 from just 35, in the first update to the rules since the 1950s (Radio Free Asia, 2013c). At the time the regulations were released, reports showed that only two Chinese cities had passed the new clean drinking water tests. Radio Free Asia (2013c) conveyed that in 2013 activists in the eastern city of Nanjing demanded that the government provide updates, and make public the details revolving around the new drinking water quality standards. Chen Yuan, leader of the civil action group, Nanjing Citizens Under Heaven, stated the reason for the protests: "The thing we really wanted to gain from this was to establish the right of the public to know what is going on ... we need to know if there is a problem, and if so, where the problems are" (Radio Free Asia, 2013c).

References

Burkholder, R. (2005, March 1) Environment pays a price for China's economic boom. Gallup. Retrieved from http://www.gallup.com/poll/15034/environment-pays-price-chinas-economic-boom.aspx

Cao, W., Han, J., Yuan, Y., Xu, Z., Yang, S. & He, W. (2015) Drinking water: a risk factor for high incidence of esophageal cancer in Anyang, China. *Environmental Geochemistry and Health*, 1–10. doi: 10.1007/s10653-015-9760-6

Carlton, E. J., Liang, S., McDowell, J. Z., Li, H., Luo, W. & Remais, J. V. (2012) Regional disparities in the burden of disease attributable to unsafe water and poor sanitation in China. *Bull WHO*, 90(8), 578–587.

Datta, S. (2015) China: The human price of pollution. LensCulture. Retrieved from https://www.lensculture.com/articles/souvid-datta-china-the-human-price-of-pollution

Davison, N. (2013, March 29) Rivers of blood: the dead pigs rotting in China's water supply. *The Guardian*. Retrieved from http://www.theguardian.com/world/2013/mar/29/dead-pigs-china-water-supply

Delang, C. O. (2016) *China's Air Pollution*. London: Routledge Focus.

Duan, W. (2013, May 20) Guangzhou finds cadmium-tainted rice. Retrieved from http://www.globaltimes.cn/content/782736.shtml

The Economist (2013, March 12) A Bay of Pigs moment. Retrieved from http://www.economist.com/blogs/analects/2013/03/water-pollution

Eiferman, P. (2014) Inside China's cancer villages: Q&A with Souvid Datta. Retrieved from http://roadsandkingdoms.com/2014/inside-chinas-cancer-villa ges-qa-with-souvid-datta/

Garland, M. (2013, March 26) China's deadly water problem. *South China Morning Post*. Retrieved from http://www.scmp.com/comment/insight-op inion/article/1199574/chinas-deadly-water-problem

Huang, Y. (2013, June 6) China: the dark side of growth. *Yale Global Online*. Retrieved from http://yaleglobal.yale.edu/content/china-dark-side-growth.

Huang, Y. (2014, July 16) The 2008 Milk Scandal Revisited. Forbes. Retrieved from http://www.forbes.com/sites/yanzhonghuang/2014/07/16/the-2008-m ilk-scandal-revisited/

Hui, Z. (2015) Antibiotic overload. *Global Times*. Retrieved from http://www. globaltimes.cn/content/928252.shtml

Ivanova, N. (2013, January 18) Toxic water: across much of China, huge harvests irrigated with industrial and agricultural runoff. Circle of Blue. Retrieved from http://www.circleofblue.org/waternews/2013/world/toxic-wa ter-across-much-of-china-huge-harvests-irrigated-with-industrial-and-agri cultural-runoff/

Kaiman, J. (2013a, February 21) Chinese environment official challenged to swim in polluted river. *The Guardian*. Retrieved from http://www.theguardia n.com/environment/2013/feb/21/chinese-official-swim-polluted-river

Kaiman, J. (2013b, May 3) China arrests 900 in fake meat scandal. *The Guardian*. Retrieved from http://www.theguardian.com/world/2013/may/03/china -arrests-fake-meat-scandal

Kao, E. (2013b, May 23) Wenzhou policeman gets sick saving girl from polluted river. *South China Morning Post*. Retrieved from http://www.scmp. com/news/china/article/1244191/wenzhou-policeman-gets-sick-saving-girl-p olluted-river

Li, J. (2015, February 1) Ways forward from China's waste problem. The Nature of Cities. Retrieved from http://www.thenatureofcities.com/2015/02/ 01/ways-forward-from-chinas-urban-waste-problem/

Liu, L. (2010) Made in China: Cancer villages. *Environment: Science and Policy for Sustainable Development*, 52(2), 8–21.

Lu, Y., Song, S.Wang, R., Liu, Z., Meng, J., Sweetman, A. J., Jenkins, A., Ferrier, R. C., Li, H., Luo, W., Wang, T. (2014) Impacts of soil and water pollution on food safety and health risks in China. *Environment International*, 77, 5–15.

Luo, C. (2013a, February 20) New offer made to another official to swim in polluted river. *South China Morning Post*. Retrieved from http://www.scmp. com/news/china/article/1154588/new-offer-made-another-official-swim -polluted-river

Luo, C. (2013b, February 26) Villager assaulted after challenging official to swim in polluted river. *South China Morning Post*. Retrieved from http://

www.scmp.com/news/china/article/1159017/villager-assaulted-after-challen
ging-official-swim-polluted-river

Luo, C. (2013d, April 10) Dongchuan district chief vows to investigate "milk
river" pollution. *South China Morning Post.* Retrieved from http://www.
scmp.com/news/china/article/1211601/dongchuan-dis
trict-chief-vows-investigate-milk-river-pollution

Radio Free Asia (2013a) Fish die from China's Yangtze pollution. Retrieved
from http://www.rfa.org/english/news/china/fish-03282013183834.html

Radio Free Asia (2013b) Toxic rice highlights China's lack of openness on pollution. Retrieved from http://www.rfa.org/english/news/china/toxic-052420131400
34.html

Radio Free Asia (2013c) Call for transparency over china's water quality.
Retrieved from http://www.rfa.org/english/news/china/quality-07012013163
443.html

Radio Free Asia (2015) Massive antibiotic pollution in china's rivers 'fueled by
abuse'. Retrieved from http://www.rfa.org/english/news/china/pollution-070
72015112452.html

Song, S. (2013) China's gutter oil scandal: 1/10 of China's cooking oil may be
recycled from garbage. *International Business Times.* Retrieved from http://
www.ibtimes.com/chinas-gutter-oil-scandal-110-chinas-cooking-oil-ma
y-be-recycled-garbage-1448384

Wang, J. (2004, October 19) Riverside villages count cancer cases. *China Daily.*
Retrieved from http://www.chinadaily.com.cn/english/doc/2004-10/19/con
tent_383720.htm

Watts, J. (2011) Exploding watermelons put spotlight on Chinese farming
practices. *The Guardian.* http://www.theguardian.com/world/2011/may/17/
exploding-watermelons-chinese-farming

WHO (1990) *Public Health Impact of Pesticides Used in Agriculture.* Geneva:
World Health Organization.

Zhang, Q. Q., Ying, G. G., Pan, C. G., Liu, Y. S. & Zhao, J. L. (2015) A
comprehensive evaluation of antibiotics emission and fate in the river basins
of China: Source analysis, multimedia modelling, and linkage to bacterial
resistance. *Environmental Science & Technology.*

Zhu, Y. G., Johnson, T. A., Su, J. Q., Qiao, M., Guo, G. X., Stedtfeld, R. D.,
Hashsham, S. A. & Tiedje, J. M. (2013) Diverse and abundant antibiotic
resistance genes in Chinese swine farms. *Proceedings of the National Academy
of Sciences* 110(9), 3435–3440.

5 The solutions

The action plan for water pollution prevention and control

In June 2014, the Ministry of Environmental Protection (MEP) released a report which stated that close to 60 per cent of groundwater monitored by 4,778 stations was rated as bad or very bad (Yin, 2015). Forty per cent of the water in the seven major river systems was found polluted and unsafe to drink. Furthermore, 17 of 31 major freshwater lakes had varying levels of pollution, "including China's two biggest lakes, Poyang and Dongting, both of which have also shrunk significantly compared to the area they covered at their recorded peaks". Moreover, in 2014 China Central Television aired a special programme on water contamination and reported that sample tests of the seven major waterways showed traces of antibiotics, with the Pearl River being the most severely polluted.

The growing public denouncements of China's water pollution have led to the country tackling the problem with a more determined effort. In April 2015, the State Council finally addressed the ever-growing public disenchantment over the worsening water pollution, with the publication of the Action Plan for Water Pollution Prevention and Control (China Water Risk, 2015c). The drafting of the plan took approximately two years and underwent 30 revisions, following in the footsteps of a similar version published in 2013 targeting air pollution. China Water Risk reports that the new plan was developed through the coordination and inputs from more than 12 ministries and government departments (Kong, 2015).

The goal of this plan is to reduce pollutants, improve drinking water, and promote water conservation. According to Bourdeau, Fulton, and Fraker (2015), the ambitious plan seeks to reverse the deterioration of water quality and vastly improve the management of water resources throughout the country. According to Yin (2015), the new plan was compiled after the Central Government vowed in 2014 to spend over CNY 70 billion ($11.42 billion) to implement a clean water action plan. "Experts say the new comprehensive plan has created a blueprint to alleviate the worsening water problems" (Yin, 2015). Bourdeau, Fulton, and Fraker state that the plan "sets progressive goals over the next 5, 15, and 35 years, and provides a wide-ranging policy agenda that includes a stricter regulation of industry effluent discharges combined with, among other things, market-based incentives, investment in new water treatment facilities, and promotion of more efficient and cleaner technologies" (Bourdeau, Fulton, and Fraker, 2015). According to the plan, by 2020, 70 per cent of the country's seven major rivers and 93 per cent of the drinking water sources in prefecture-level cities should meet acceptable standards. By the end of 2030, the figures should rise to 75 per cent and 95 per cent, respectively. By 2050, the overall improvement of water quality nationwide is presumed. Bourdeau, Fulton, and Fraker write that the plan calls for a reduction in the prevalence of "black and odorous water bodies" to less than 10 per cent by 2020, and to be completely eliminated by 2030.

According to Wang Dong, a research fellow at the Chinese Academy for Environmental Planning, "The government's efforts to combat water pollution have been focused on major river valleys. The new plan has more comprehensive and systematic coverage with the inclusion of small streams". Mu Jianxin, a senior engineer from the Department of Irrigation and Drainage at the China Institute of Water Resources and Hydropower Research, adds: "The goals set out in the plan are ambitious and require prolonged efforts, particularly where it concerns heavy metal pollutants which are difficult to remove from water" (Yin, 2015).

The plan is grouped into the following ten programmatic areas: "Pollutant Discharge Control; Economic Restructuring; Conservation and Protection of Water Resources; Scientific and Technological Support; Market Mechanisms; Environmental Regulation and Enforcement; Environmental Management; Protection of the Aquatic Ecological Environment; Differentiated Responsibilities; and Public Participation and Social Supervision" (Bourdeau, Fulton, and Fraker, 2015). Within the "Pollutant Discharge Control" programme there are four major components of interest.

1   *Industrial production.* The focus is on establishing cleaner standards for the ten major industries – papermaking, nitrogen-based fertilisers, steel, non-ferrous metals, textiles, agricultural products, pharmaceuticals, leather tanning, pesticides, and electroplating.

2   *Wastewater treatment.* The goal here is for a "nationwide infra-structural expansion of industrial and municipal wastewater treatment facilities".

3   *Agriculture.* It is expected that there will be major changes in the approach to agricultural practices, including livestock farms, which will be restricted in certain regions; some farms will be forced to relocate; treatment of sewage will be mandatory; and in areas of water scarcity there will be a requirement for a transition to be made from water-intensive to drought-tolerant crops. Moreover, regulations and standards will be set to "reduce nonpoint source pollution from agriculture by reducing the use of fertilisers and promoting or requiring less toxic and persistent pesticides".

4   *Shipping.* Measures will be provided to reduce pollution from the shipping industry with stricter standards to go into effect for new coastal vessels after 2018, and new inland river vessels after 2021. Besides additional requirements for international ships, "ports will be subjected to stronger pollution control plans and waste management and water treatment standards".

Hewitt (2015) writes that to achieve some of the plan's goals the government has pledged to shut down small and inefficient factories which lack the resources to ensure they are environmentally sound, in 10 polluting industries (including papermaking, iron and steel, pesticides, and tanning) by the end of 2016. Larger plants will be required to reduce emissions and upgrade technology. A blacklist will be prepared to delineate businesses exceeding their pollutant quota with major violators risking their business being shut down. New projects will be subjected to greater control measures. Hewitt also reports that many experts believe the implementation of environmental regulations at the local level will be a challenge for a nation that has concentrated on achieving fast economic growth for decades. Hewitt cites Chinese Premier Li Keqiang who said that getting officials to implement policy swiftly and effectively was one of his hardest tasks. Hewitt also quotes Mu Jianxin, a senior engineer at the China Institute of Water Resources, who expressed his concern that the water action plan might encounter resistance from industries and local governments concerned over potential loss in profits, and such "pillar industries" as iron and steel that might get hit hard because of the requirement to upgrade their technology, which would lead to a rise in production costs. Mu feared this could possibly stall economic development in the region. Nevertheless, the government plans to offer financial marketing incentives to industries to encourage them to be on board with the plan. Companies that exceed their pollution quota are warned they will be named on a public blacklist and serious violators will encounter closure.

Despite China's promise to spend CNY5,000 billion to improve water pollution supplies over the next decade (CDP, 2015), trusting the government to live up to its word is another issue among Chinese citizens. For example, 20,000 residents in the town of Changshou in Hunan Province's Pingjiang County have refused to drink the local tap water for years despite quality control tests purportedly showing no evidence of contamination (Kao, 2013a). Many believe the tap water is too polluted to drink, cook, or bathe in because the local river is being polluted by a gold mine upstream. The majority of the residents will only use mountain spring water for their needs regardless of the officials who steadfastly claim that the water is not contaminated and has consistently met national standards. Wu Jinan, deputy director of the Pingjiang Environmental Protection Bureau concurs with this assessment, saying that all the small businesses operating mines in the area have been shut down by 2008 and only one gold mine is still in operation. He is adamant that all pollution and emission standards have been met. Indeed, Kao writes that He Peiyu from the Chinese Academy of Social Sciences believes the fears over tap water are simply a trust issue. In an interview with the *People's Daily* he states, "People are not convinced, and there are no news channels between officials and the people. The government needs to actively work on reconstructing its own credibility problems, clarify its intentions and identify counter-measures".

Integration of management

Yu (2011) reports that on 29 January 2011 the Chinese government declared in its No 1. Document – the central government's first policy document of the year, setting the top priorities – that it plans to invest CNY 4 trillion over the next ten years to protect and improve access to water. The water strategy was announced during one of northern China's most severe droughts. Some of the measures proposed in the document include "control of total water consumption, improved irrigation efficiency, restricted groundwater pumping, reduced water pollution, and guaranteed funds for water-conservancy projects". Though the high priority placed on a sustainable water use plan is welcomed, and is a step in the right direction, Yu argues that it will be difficult to put into practice.

Yu (2011) writes that since the 1950s, China has constructed 86,000 reservoirs, drilled over 4 million wells, and developed 58 million hectares of irrigated land, which has generated 70 per cent of the country's total grain production. But the conservation of water has lagged behind. Yu calls the threat to sustainable water supplies "a growing geographical mismatch between agricultural development and water resources":

> The centre of grain production in China has moved from the humid south to the water-scarce north over the past 30 years, as southern cropland is built on and more land is irrigated further north. As the north has become drier, increased food production there has largely relied on unsustainable overuse of local water resources, especially ground water. Wasteful irrigation infrastructure, poorly managed water use, as well as fast industrialization and urbanization, have led to serious depletion of groundwater aquifers, loss of natural habitats and water pollution.
>
> Yu, 2011, p. 307

Yu argues that one problem that still needs to be addressed in the document is the scattering of authority across different agencies. One example is that major rivers are managed by the Ministry of Water Resources, while the local governments control smaller water courses. Not only are they administered on a separate basis, but data on such things as water supply, farmland irrigation, weather forecasting, river runoff, groundwater, land use, water pollution, water use, and more, are not shared between the various agencies or made transparent to the public. Yu stresses, "It will be difficult to implement the holistic policy laid out in the No 1. Document without breaking down these bureaucratic barriers".

Among the numerous issues that require more attention in the document, Yu contends, "China needs to build an integrated network to monitor surface and groundwater, and use it to assess and set water policies through an integrated water-resource management system" (Yu, 2011). Plus, there is a need for legislation setting clear policies on data sharing, and penalties for those who fail to comply. Yu also writes that as political awareness about water issues increases, a new, fairer water law is essential to protect citizens' rights and prevent corruption. For example, if water prices rise, low-income farmers will be severely affected – "to protect them, and so food supplies, China must keep irrigation costs low" (Yu, 2011).

Another area in the document that Yu identifies as being vague are the guidelines of how the departments of water resources and environment protection should cooperate on water supply and the problem of water pollution. Although the document mentions such issues as ecological water use, no measures are specified for how to protect the needs for water of ecosystems against the conflicting demands for water of economic activities.

Appointment of river administrators

One approach used to address the scattering of authority across different agencies lamented by Yu (2011) is the appointment of river administrators. To ensure better water quality, river administrators have been appointed since 2008 to oversee pollution control of a number of waterways. This practice has proven to be effective in improving river health and has gradually been adopted by many other places throughout China.

According to Xia, Zhang, Zhan, and Ye (2011), "over the last 30 years or more China has established an extensive water pollution-control system with a large set of institutions, a variety of legislation and policy instruments, and comprehensive investment plans that were largely shaped within the traditional five-year plan preparation process. In the earlier phases water pollution control was characterised mostly by command-and-control instruments (industrial permit systems, simultaneous control programmes) and partly by economic instruments (such as pollution levy fees and discharge permits)" (p. 169). The diversity of the approaches and the large set of institutions makes river management more difficult. Appointing a river administrator to overlook or take over the responsibility for managing the river under his or her administration addresses the problem of having a multitude of organisations, perhaps with different objectives and using different tools or approaches, and makes the management of the rivers more straight-forward.

Sun Jichang, Director of the Construction and Management Department of the Ministry of Water Resources states: "The practice of appointing official river administrators aims to encourage local governments to integrate resources and take full responsibility for the ecological quality of rivers and lakes" (Wang, 2014).

Earlier in this volume, I discussed the plight of the Tai Lake. Wang (2014) cites state broadcaster CCTV's report that the algae built up near one of Wuxi City's major water plants made the city's water putrid and undrinkable for nearly 2 million residents, and discusses some of the solutions offered to combat this problem and clean up the Tai Lake. Wang details that after the drinking water crisis a number of emergency measures were employed. These included "removing the blue-green algae, sterilizing water with activated carbon, transferring water from the Yangtze River into the Taihu Lake, and artificial rainmaking to dilute pollution". However, further steps towards the long-term management of the river were also taken. In August 2008, a document was issued by the city making river water quality management an important criteria in selecting officials and evaluating their performances. Subsequently, Wuxi's officials were appointed administrators of major river sections under their jurisdiction. Pleased with the administrator's progress, the Jiangsu Provincial Government promoted the Wuxi-style river management across the entire province, and leaders were appointed to coordinate pollution control efforts in 15 major rivers.

The Vice Mayor of Wuxi, Zhu Aixun, heartily approved the reforms, stating that the leadership of river administrators has allowed strong measures to be taken to control water pollution, including "ecological rehabilitation, shoreline adjustment, sealing sewage draining exits along rivers, and removing sludge on riverbeds". Zhu Aixun added that water quality has improved in all river sections of the city, and drinking water has met national quality standards. Wang (2014) noted that Zhang Lijun, formerly Vice Minister of the State Environmental Protection Administration (now the Ministry of Environmental Protection), acknowledged the merit of the river administrator system, but worried that since environmental protection is only one of the many responsibilities of local officials, they may put their responsibility for growing the economy above that of environmental protection. According to Wang Canfa, a law professor at China University of Political Science and Law in Beijing, "a fatal flaw in the river administrator system is that its results hinge on the amount of attention that officials attach to pollution control". Canfa suggests that environmental responsibilities of local officials should be written into the law and their performance should be evaluated based on well-defined criteria and procedures. Wang also quotes other experts who state that although the river administrator system may be effective in curbing pollution in small local rivers, cooperation between local governments for pollution control in cross-regional rivers is essential.

Supply management vs. demand management

The challenges of insufficient water resources and degraded water quality caused by widespread pollution have forced China to be engaged in a delicate balancing act. Cheng and Hu (2012) contend that "China faces an imbalance between the supply and the demand of water for supporting the rapid social and economic development while protecting the natural environment and ecosystems" (p. 253). They anticipate that climate change will stress freshwater resources even more, and widen the gap between the demand for, and supply of water. The researchers define the term "water resources development and management" as "the actions required to manage and control freshwaters to meet human and environmental needs" (p. 263). The researchers provide a historical perspective on China's water resources management policies and practices; and strongly recommend "demand management and pollution control as key measures for improving water resources management to adapt to climate change based on the current political, socioeconomic and water resources conditions" (p. 253).

Cheng and Hu (2012) write that many water supply sources in China (including rivers, lakes, and groundwater basins) are over-exploited and heavily polluted, and often unable to sustain natural aquatic ecosystems. They add that "climate change has already aggravated and may further aggravate the water challenge in China, worsening the water shortages and intensifying the conflicts among water users" (p. 256). In light of this, the researchers assert that "adaptation measures in water resources management policies and practices are vitally necessary to meet the challenges and upcoming climate change. And this applies to the use of both supply-side and demand-side adaptation strategies to ensure water supply during average and drought conditions" (p. 256). The authors proclaim that:

> China's history has shown that its water resources policies and management practices have proven to be largely supply-driven, engineering-based, single-sector and usually involve only a single stakeholder. Intensive efforts have been placed on the assessment and development of new water sources and installation of delivery systems to meet socio-economic needs for water. Such supply-driven water resources management largely failed to account for the economic nature of water resources in relation to their natural characteristics and aggravated the natural water shortages. Little attention was put on exploring efficiency and demand management solutions to reduce the wasteful water consumption until recently.
>
> p. 271

After thoroughly assessing the present state of socioeconomic and water resources, Cheng and Hu (2012) emphasise that China should adopt an emphasis on "demand management as the primary strategy while controlling water pollution to quench the increasing water demands from the rapid social and economic growth and to adapt to climate change. Important issues such as water use efficiency, water rights and water rights trade, and effective enforcement of laws and regulations should be addressed while switching to demand management" (p. 279). However, the authors do propose that both supply and demand-side management options should be utilised in the development of a sustainable water resources management strategy to make the limited water supplies meet the demands of economic development, social well-being and the conservation of ecosystems in the context of global climate change.

Increasing the price of irrigation water

Fanus and Zhao (2012) argue that the combination of irrigation and chemical fertiliser plays an indispensable role in China's rapid growth in agricultural output. However, the inefficient usages practiced by the farmers have exposed the country to environmental and water pollution, including the increased production costs arising from the overuse of fertiliser, coupled with a decreasing soil response to fertiliser application. Farmers are being blamed for contributing to pollution problems and not taking enough, if any steps to reduce it. To address this problem, the Chinese government has proposed integrated water management plans, which includes increasing irrigation water price. The water price has been used as a leverage under the market mechanism to promote the national conservation and allocation of water, since the under-pricing of irrigation water may lead to the excessive use of water.

The policy is not without its controversy. One fear is that increasing the price of irrigation water will prompt farmers to use less water and more artificial fertilisers, which as mentioned above may lead to the drying of the soil and increased water pollution. However, Fanus and Zhao found the opposite. They set out to analyse the relationship between irrigation use and fertiliser use decision, and based on the primary data from 87 farm plots along the Weihe River basin (Shaanxi Province) they found that irrigation use and inorganic fertiliser application are highly correlated. They concluded that irrigation water use was a significant determinant factor of the fertiliser use decision of farmers, and that policies that improve irrigation water efficiency can potentially "positively influence fertiliser use efficiency" (p. 80).

Other problems have been identified by Zhou, Wu, and Zhang (2015), based on a fieldwork in Zhangye City (in arid Gansu Province). Their conclusions are that the increase of irrigation water price has little impact on water saving, because even after the increase in the irrigation water price, water is still cheap. Farmers are reluctant to adopt new technologies to save water. Furthermore, many farmers use groundwater, which is almost free, and the quantity extracted is difficult to monitor. They conclude that the "increasing surface irrigation water price may lead to a further overexploitation of groundwater, considering the low price of groundwater and the unrestricted extraction of groundwater on existing wells and the not effectively enforced control on digging new wells" (p. 6). According to Zhou et al. (2015), the number of wells has increased by more than 25 per cent, and groundwater use for irrigation by more than 70 per cent after the price of water for irrigation increased in Zhangye.

The additional use of groundwater by farmers may also have indirect, unexpected, and detrimental consequences on the quality of farmland and food security. Gao, Yu, Luo, and Zhou (2012) report that the groundwater quality in many cities is low (Class III or IV of the national water standard) because the water includes NO_3-N that poses risks to human health. The water is unfit for human consumption, and using the low quality groundwater to irrigate may not only affect fertiliser use efficiency, but would also enhance the non-carcinogenic risk for people who consume it (maybe indirectly through food consumption), or come in contact with the water, including during the irrigation process (Gao et al., 2012).

Wastewater discharge

The Chinese Ministry of Land and Resources launched a 6-year investigation in 2006 to assess the scope of the problem of groundwater contamination from untreated wastewater discharge on the North China Plain, the region most dependent on groundwater and home to nearly 130 million people. Subsequently, in April 2013, the government presented a highly ambitious work plan for controlling groundwater contamination in the area. Li (2013) states that the extent of the problem is unclear because the full details of the 2006 survey were not made public for fear of alarming people. The only thing the government plan acknowledged was that the pollution levels were very serious. The government said it will divide the North China Plain into 30 units for pollution prevention and control, into three severity categories – serious, poor, and good. The plan will include an investment of CNY 500 million between 2013 and 2020 to beef up environmental regulation, and a number of measures across the country including "to increase pollution assessments and establish a database of results; to control river pollution from agriculture and point sources from industry and landfill; to treat polluted areas; and to conduct more research into clean-up and prevention strategies" (Li, 2013, p.15). Lastly, researchers will examine the effects of shale gas development on groundwater.

South-to-North Water Diversion Project (SNWDP)

As mentioned above, China experiences an important water imbalance with a surplus of water in the south, and water scarcity in the north. This imbalance has always existed in China, and leaders have been calling for the construction of canals to address the regional imbalance for a very long time. For example, Mao Zedong has been quoted as saying "Water in the south is abundant, water in the north scarce. If possible, it would be fine to borrow a little" (in Dillow, 2014), but the idea was first raised well before his time. Finally, on 27 December 2002, China officially started the construction of the South-to-North Water Diversion Project (SNWDP) in an effort to meet the water and energy demands of urban centres, industries, and agriculture in the dry northern and western provinces. The whole project was expected to cost CNY372 billion, although in 2014 the budget was raised to CNY486 billion (Chang, 2014), becoming the most expensive civil engineering initiatives in history. It is expected to be completed in 2050.

This construction will link the Yangtze River watershed in the south to the Yellow, Huaihe, and Haihe rivers in the north, and divert 44.8 billion cubic meters of water yearly via 4,345 kilometres of canals, pipeline, aqueducts and pumping stations. Dillow (2014) describes it as China's ongoing push to engineer around a growing water crisis without interrupting economic development. Three lines are planned (Figure 5.1): the East Line, which will divert the water from the lower Yangtze northward to Tianjin, extending 1,150 km in length with a designed discharge of 800–1000 m^3/s; the Central Line, which will extend 1,241 km in length from the upper Hanjiang River to Beijing and Tianjin with designed discharges of 630 m^3/s; and the West Line, which will divert the water from the upper Yangtze to the upper Yellow River (Chen, Zhang, and Zhang, 2002).

Dillow (2014) disclosed that engineers and water resources experts are concerned that projects as sprawling as the SNWDP may "create new, long-term risks the project petitioners won't be able to mitigate or project sponsors don't care to address" (p. 8). One such concern is the water quality. Some media (e.g. Liao, 2013; Jing, 2014; Lin, 2014a) are already reporting that the pollution brought by the Yangtze River water being pumped in other waterways is threatening the livelihood of farmers and fishermen living along the route. Since the water quality

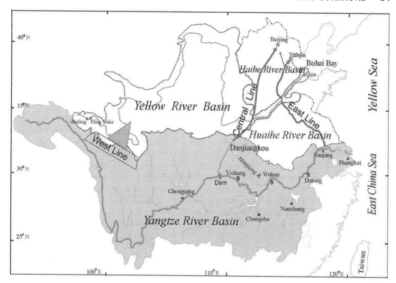

Figure 5.1 Geographical settings of the Yangtze and the other three large river basins in North China and a sketch map of the SNWDP as indicated by the West Line, the Central Line and the East Line
Chen et al., 2002

worsens as the Yangtze River flows downstream, of particular concerns are the water bodies where the Central and East Lines flow in on their way north. Several studies focus on water pollution on the Central Line of the SNWDP. The Central Line is of particular concern because it diverts water from the Danjiangkou Reservoir in Hubei Province to Beijing's Tuancheng Lake, for consumption by households in Beijing and the surrounding region.

According to Tang, Yi, Yang, and Cheng (2014), the Central Line "is composed of a long canal and complex hydraulic structures and will transfer water in open channel areas to provide drinking water for Beijing, Shijiazhuang and other cities under extremely strict water quality requirements" (p. 2111). However, they argued that "canals from the southwest to the northeast cut through original roads; as a result, 27 highway bridges were constructed to allow normal traffic" (p. 2115), and that "a large number of vehicular accidents [...] on the many highway bridges across the main canal would cause significant water pollution in the main canal" (p. 2111).

Ma et al. (2014) looked at the pollution in the water sources of the Central Line (the Danjiangkou Reservoir). They measured five water quality indicators (dissolved oxygen, permanganate index, ammonia nitrogen, total nitrogen, and total phosphorus) at three monitoring sites (the Danjiangkou Reservoir dam, the Hejiawan, and the Jiangbei bridge), "to investigate changing trends, and spatiotemporal characteristics of water quality in the Danjiangkou Reservoir area from January 2006 to May 2012" (p. 242). Their results are that:

- the concentrations of all pollutants increased from 2006 to 2012, with the concentrations of phosphorus and nitrogen being particularly worrisome;
- the seasonal concentration of the five water quality indicators varied with changes in temperature, rainfall and runoff. Overall, the water quality of the Danjiangkou Reservoir was worse during the wet season than during the dry season;
- water quality evaluation results indicated that the probability of the water quality belonging to type II (lightly polluted, but can be used for drinking after treatment) was 27.7–33.7 per cent.

To reduce the concentrations of nitrogen and phosphorus, they suggested implementing integrated water quality management strategies to avoid the eutrophication of the Danjiangkou Reservoir. "These strategies should include the control of man-made pollution sources, control of fertiliser use on upstream farmlands, and increasing the vegetation coverage on the upper regions of the reservoir to prevent water loss and soil erosion" (p. 249). Thus, once the SNWDP is complete, further management efforts will be needed to address the additional pollution in those water bodies that are still relatively clean.

Dredging

One other method used to mitigate the impact of water pollution is environmental dredging. Zhang, Zeng, Liu, Zeng, and Jiang (2014) view environmental dredging as a very significant and efficient means of counteracting the eutrophication of water bodies caused by the endogenous release of nitrogen and/or phosphorus from polluted sediments. Tremendous amounts of anthropogenic pollutants rich in nitrogen and phosphorus have been discharged into waterways, leading to serious eutrophication problems. Zhang et al. (2014) stress that the treatment or disposal of polluted sediments is necessary to eradicate the eutrophication of water bodies and prevent worsening conditions in the future. "The pollutants in water can be absorbed by sediment and accumulated to a high concentration. When the hydraulic conditions (e.g., temperature, flow velocity, perturbation, and pH) change, the accumulated pollutants in the sediment may be released into the water body, leading to a secondary pollution" (p. 1119).

The researchers write that despite the efforts made over recent years to control and diminish exogenous pollutants in bodies of water, eutrophication still exists due to the release of pollutants from sediment. Sediment dredging consists in removing pollutants from sediments as an alternative to the treatment of such sediments, and is widely used to restore a body of water. The researchers argue that the "huge operational cost and subsequent disposal cost of the dredged polluted sediments, as well as the adverse effect on the benthic environment caused by excessive dredging, make the current dredging method [unsuitable]" (p. 119). "Most benthic communities and fish need a certain amount of organic sediment to keep alive. Excessive dredging not only damages the benthic environment, but also increases the dredging costs. [...] It is not advisable to dredge all the sediment from a water body" (Zhang et al., 2014, 1120).

With regards to the vulnerability of the ecological environment and economic factors, Zhang et al. (2014) propose the concept of precise dredging, in which only a certain depth of sediment containing the highest concentration/level of pollutants is dredged, leaving a portion of sediments with nutrient properties for the benthic community. Compared to the traditional dredging method, precise dredging was noted to save 16 to 45 per cent of the cost. The Nanfei River was chosen as a case study of how the concept of precise dredging could be realised. The river flows through Hefei city in Southeast China and drains into Chaohu Lake, one of the five biggest freshwater lakes in the country. The Nanfei River is also one of the most heavily polluted rivers because of rapid expansion and overpopulation. This study proved that the precise dredging method dredged a smaller amount of sediment, but nevertheless suitably removed pollutants including heavy metals and persistent organic pollutants. As a result of this work, the precise dredging method was later adopted by the National Water Project of China to treat the endogenous pollution of the Nanfei River in 2010.

New water standards

As mentioned in Chapter 4, another step towards improving China's water situation came on 1 July 2012, when the country's new standards for drinking water quality officially took effect. In the first update to the rules since the 1950s, the number of items to be tested increased from 35 to 106. Qu, Zheng, Wang, and Wang (2012) reveal that all 106 items included were adopted from the WHO guidelines for drinking water quality. Over the years, there had been some revisions of the standards, but these were considered "soft targets", rather than "hard requirements". For instance, according to He and Charlet (2013), in 2006, the Ministry of Health revised the standards of drinking water quality "with permitted arsenic concentration as 0.01 mg/litre instead of 0.05 mg/litre as used in the past decades" (p. 83). However, this level will only be strictly enforced with the new standard, suggesting that more people "will be protected if they have been exposed to drinking water with arsenic concentration of 0.01–0.05 mg/L" (p. 83).

The new water standards are considered a milestone in regards to China's environmental legislation; it signals a vigorous and concerted effort to improve drinking water quality and public health. After these reforms, China's number of regulated items exceeded those of many developed countries. According to Qu et al. (2012), "This is the first time a developing county has implemented strict regulations on drinking water quality, and, in China, the first time the same standards have been applied in rural and urban areas". However, the authors outline that obstacles such as severely polluted water sources, outdated water-processing facilities and techniques, and low investments to improve water plants by provincial and municipal governments make it challenging for China to meet the new standards. Improving water quality in rural areas is identified as especially complex because these areas lack sufficient sanitation infrastructure to handle waste disposal, sewage collection, and water treatment. Qu et al. highlight that China's rapid industrial development and the relocation of factories from urban to rural areas worsen the circumstances, and are essentially at odds with regulation efforts. A shortage of staff and technology also presents a problem because only the Centers for Disease Control and Prevention at the provincial level are able to properly monitor all the items listed in the standards. These obstacles will make it more difficult to implement the standards.

Qu et al. (2012) also express concern as to how well the regulations fit China's situation. The researchers cite a spectrum analysis of drinking water from the Huangpu River in Shanghai, where more than 200 pollutants were present: "50 per cent of the items specified in the standards were not detected in drinking water, yet some samples contained high-risk compounds which are not included in the standards". The authors conclude that China should reassess the new standards and "monitor their implementation to ensure the values and items specified are appropriate for the nation, rather than simply matching international standards".

Household water purification

Tao and Xin (2014) agree that making drinking water safe is a priority in China. However, the authors contend that tackling pollution and using different grades of water for different tasks is much more efficient than making all water potable. Tao and Xin discuss the 2012 mandate that tap water in all of China's cities should meet the standard based on 106 indices, and how billions have already been spent on researching drinking water problems in key river basins and lakes – but they also note that only a few cities met the desired standard. Calling it a problem of supply and demand, the authors proclaim that "supply is a challenge because almost half of China's water sources are polluted", and water demand is a challenge because of the "runaway economic growth and urbanization" (p. 528).

The authors mention that the Chinese government is in the middle of a CNY 410 billion five-year programme to deliver safe drinking water to all town and city residents – about 54 per cent of the population – by 2015, with a focus on "upgrading 92,300 kilometres of main pipes and thousands of water treatment plants to developed-world standards". Tao and Xin (2014) feel this development of infrastructure is ill-suited to China, especially since China is projected to remain a developing country until around 2050. According to Tao and Xin, bringing a large developing country like China up to the same standard as a developed country will require very intensive water treatment, which may have negative environmental consequences. They write, "Urban expansion will outpace improvements to public water systems, and treating polluted water will require large amounts of energy, expensive technologies and chemicals". They provide the example of Jiangsu Province, where "carbon dioxide emissions increased by 28 per cent in 2012 when a type of water filtration called ozone-biological activated carbon treatment was extended to one-quarter of the province's supply (5.3 million tonnes per day). China is in need of cheap, energy-efficient methods of water purification that minimize chemical use" (p. 528).

Their solution to the water problem issues is to initially focus on cleaning the water sources and recycling water. The researchers suggest the first course of action should be to purge rivers and lakes of industrial and agricultural pollutants and prevent them from entering the water table. They point to using cheaper technologies such as purifiers on taps, and claim that it would suffice for delivering clean drinking water to most of the country's population "because drinking water accounts for only a few per cent of total consumption. Water of lower quality can suffice for laundry, bathing, and kitchen use" (p. 527). The researchers call for purification systems that improve drinking water at the point of use. They draw comparisons with Kenya, Bolivia and Zambia, where water purifiers have proven to reduce diarrhoeal disease by 30–40 per cent. Currently, only 5 per cent of Chinese homes have water purifiers, "despite a unit costing only around CNY 1,500 to 2,000". China's water-purification industry is growing by about 40 per cent a year, but the authors warn that water-purification devices are unregulated. They also point out that treated grey water (wastewater from showers and baths) and black water (from toilets) are frequently used for industrial and irrigation purposes, and for flushing in new residences; but they consider this type of recycling impractical for most households because of the high costs and the difficulty of installing the necessary type of plumbing. "By using cheap, low-carbon water purifiers in all homes, China can avoid the technology 'lock-in' that leads developed countries to waste potable water, and leap-frog to a sustainable supply system. In the long term, the improvement of water sources will ensure that most people have safe drinking water" (Tao and Xin, 2014, 528).

Zhang (2012) set out to assess the impact of a drinking water infrastructure programme that began in the 1980s, which aimed to build water-purification plants and pipeline systems to address pollution problems and improve water quality and the health status of adults and children in rural China. Their research was based on a survey carried out by the China Health and Nutrition Survey (CHNS) which includes approximately 4,500 rural households in 152 villages from 1989 to 2006. The results showed that the water quality improvement resulting from this programme has generated positive and significant effects on health. The illness incidence of adults decreased by 11 per cent and their weight-for-height ratio increased by 0.835 kg/m. Children's weight-for-height ratio rose by 0.446 kg/m and height by 0.962 cm after the programme was launched. Given the fact that some villages were surveyed only shortly after the implementation of the programme, the long-term benefits should be even greater. As the author points out, "the results clearly indicate that the construction and implementation of water plants in rural China have resulted in health benefits for adults and children" (p. 132). However, they also acknowledge that "to the extent that these water treatment plants are costly to construct, and to the extent that we are still only able to see short-run benefits on health, a full analysis of the health benefits awaits future research" (p. 132).

Closure of polluting factories

Included in the new action plan to protect the country's water scarcity is the intent to ban water-polluting paper mills, oil refineries, pesticide producers, and other industrial plants by the end of 2016. According to Chen Jining, Minister for Environmental Protection (MEP), China has already blocked the approval of 165 state-level projects during the 12th five-year plan period (2011–2015) (Patton, 2015). However, many companies will ignore environmental impact assessments or find ways around the requirement. In addition, it is cheaper for companies to pay fines, which are set very low, than to meet regulations. Chen also remarks that local governments are "more focused on growth and generating tax income than environmental protection", so they dislike closing companies that contribute to the local economy. Patton mentioned that according to a nationwide MEP survey, during the first half of 2015 around 30,000 companies constructed projects illegally. Wang called for a revision of environmental laws to increase the penalties for violations.

One case where the fine was rather steep is that of the U.S.-based J. R. Simplot's China processing unit, which supplies frozen French fries to McDonald's Corp. Wong and Jourdan (2015) reported that a local regulator in Fengtai district of Beijing discovered that Simplot Food Processing Co. Ltd. had been discharging contaminated wastewater that exceeded acceptable levels. In April 2015, Beijing authorities fined the company CNY 3.92 million for water pollution. A McDonald's spokesperson said that it was taking the infraction seriously and had the assurance of Simplot that they have implemented a corrective action plan, and that they would be holding them "accountable for implementation and enhanced procedures for compliance".

Wang, Webber, et al. (2008) point out that "about two-thirds of the total waste discharge into rivers, lakes and the sea derives from industry, and about 80 per cent of that is untreated. Most of the untreated discharge comes from rural industries" (p. 648). However, Hu and Cheng (2013) point towards some success towards reducing wastewater generation by industry. According to MEP (2012), the intensity of wastewater generation and the discharges of chemical oxygen demand (COD) and NH_4^+-N from industrial sources has been declining steadily over the last decade. According to Hu and Cheng (2013), this resulted from "increased water use efficiency and better wastewater treatment, as well as improvement in industrial processes and forced closure of some of the heavily polluting factories" (p. 66). Figure 5.2 presents the COD discharge intensity from four heavily polluting industries. The COD discharges per unit of GDP from all these industries decreased exponentially during this period (Hu and Cheng, 2013).

With regards to addressing the urban waste issue affecting China, cities have attempted adopting a variety of methods to combat this

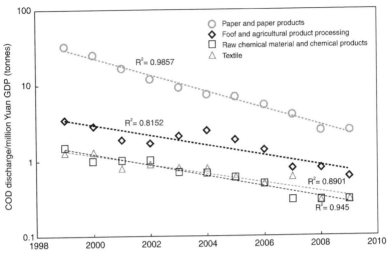

Figure 5.2 Chemical oxygen demand (COD) discharges per unit of GDP between 1999 and 2009

Hu and Cheng, 2013

problem. According to Li (2015) some cities years ago tried highly technical composting systems which were created to sort mixed waste mechanically and compost the biodegradable portion of the urban waste stream, but it was soon abandoned by the majority of users because the toxic sludge output from the composting process was unusable and became a public health hazard. Another practice that has found acceptance in the U. S. and Europe is waste incineration, which is fuelled by the idea that burning waste will address landfill space limitation issues and the energy from the incineration will generate revenue for the city. However, Li argues that the "unsorted Chinese urban waste stream, with high proportions of damp organic material does not make for efficient incinerator fuel. Much more fuel is required to burn damp waste, increasing the costs and decreasing if not nullifying the profits from energy generation". Li also contends that these waste incinerators are poorly regulated, and the resulting toxic air pollution is both an environmental and a public health issue that will impact nearby poor communities the most.

Still, Li (2015) believes the government's recent interest in using anaerobic digesters to decompose organic waste and capture the methane as a fuel source is a positive shift in the right direction, and adds that the government is currently using many large-scale anaerobic digester pilot projects in the country. In addition, Li lists a few key areas for decreasing the "environmental footprint" of China's urban waste.

- With pork being a large part of the Chinese diet, feed food scraps to locally raised pigs, which reduces the cost of pork and feed production, and transportation.
- Divert organic waste from the landfill or incineration to support managed composting and anaerobic digesters that can provide high quality natural fertiliser.
- Enhance the cooperation of the government with the informal sector to yield a more efficient, regulated, and orderly urban recycling system that will help poor migrants while generating revenue.
- Encourage local business to reuse recycled materials.

Conversion of human and animal excreta to fertiliser

According to Wu, Maurer, Wang, Xue, and Davis (1999), "most cities and communities simply discharge untreated wastewater and sewage produced by households and industries into surrounding surface waters, including rivers, lakes, and coastal areas. The increasing volume of human excreta as well as toxins and solid wastes from industries in urban areas is leading to a severe deterioration of water quality. In fact, most urban water pollution is linked to organic loads entering water bodies. In 1996, more than 20 billion tons of urban sewage was discharged into rivers, lakes, or seas, with approximately 10 per cent receiving any treatment by the 135 centralized sewage treatment plants that operated nationwide" (p. 253). Consequently, they found that by the late 1990s approximately 700 million people, over half the population in China, were consuming drinking water contaminated with levels of animal and human excreta that exceeded maximum permissible levels "by as much as 86 per cent in rural areas and 28 per cent in urban areas" (p. 251).

Jin, Zhang, and Tian (2014) look at "the current state of wastewater treatment plants (WWTPs) in urban China from the aspects of scale, treatment processes, sludge handling, geographical distribution, and discharge standards" (p. 85). They found that "wastewater treatment has developed rapidly together with the economic development in China". Jin et al. (2014) make the point that in the 12th Five-Year Plan, more than CNY 450 billion was allocated for sewage treatment. "Generally speaking, the treatment processes of China were in line with international standards. Removal of the conventional pollutants was effective, and the COD removal efficiency could reach 85 per cent on average" (p. 93). However, insufficient amount of sewage is treated, which exacerbates water-body pollution, worsening the environment and threatening human health. This is for two reasons in particular. First, people "are unwilling to pay fair prices for public services such as water and wastewater treatment" (p. 94). Second, to control the core consumer price index (CPI), which includes the price of water and wastewater treatment, "mayors hesitate to increase the price of water and wastewater service, and choose to postpone the necessary investment" (p. 94). Jin et al. among others, recommend recovering phosphorus from wastes. Phosphorus can be recovered from wastewater by precipitation as magnesium ammonium phosphate (struvite) and calcium phosphate (Driver et al., 1999). Jin et al. (2014) report that there is not yet an industrial application of phosphorus recovery in China. They also contend that previous studies reported the energy contained in sewage was 10 times larger than the energy needed for the sewage treatment. For this reason, they recommend that China recover energy from sludge as biogas through "sludge anaerobic digestion, high-efficient denitrification technologies, and chemical phosphorus removal" (p. 94).

Human and animal droppings are sometimes converted to fertiliser in rural areas in China. For instance, human waste was used as an agricultural fertiliser in 51 per cent of households on average from 2007 to 2010 in 36 villages in Sichuan Province (Carlton, Liu, Zhong, Hubbard and Spear, 2015). However, much of the human waste produced in urban areas ends up in rivers. Khan (2015) writes that in 2013 there were 731 million urban dwellers in China, overtaking the rural population by more than 100 million. The fallout from this resulted in considerable toilet waste routed into rivers. A major concern is on how to keep the waste-recycling process hygienic, especially since China may be far ahead of other developing countries in terms of urban wastewater collection systems, but not as advanced in its treatment rate. Based on a farm model originally used to keep humans from doing their business in pig troughs, now more than 40 million farm homes throughout China have a holding tank for human and animal waste that is partly sanitised by depriving the solids of oxygen; what remains is converted to liquid fertiliser for farms (Khan, 2015).

References

Bourdeau, K., Fulton, S. & Fraker, R. (2015, May 8) China announces new comprehensive water pollution control plan. Beveridge & Diamond. Retrieved from http://www.bdlaw.com/news-1734.html

Carlton, E. J., Liu, Y., Zhong, B., Hubbard, A. & Spear, R. C. (2015) Associations between Schistosomiasis and the use of human waste as an agricultural fertiliser in China. *PLoS Neglected Tropical Diseases,* 9(1), e0003444. doi: 10.1371/journal.pntd.0003444.

CDP (2015) Enhancing water security in China. Carbon Disclosure Project. Retrieved from https://www.cdp.net/CDPResults/uk-china-water-report-2015-web-english.pdf

Chang, G. G. (2014, January 8) China's water crisis made worse by policy failures. *World Affairs.* Retrieved from http://www.worldaffairsjournal.org/blog/gordon-g-chang/china%E2%80%99s-water-crisis-made-worse-policy-failures

Chen, X., Zhang, D. & Zhang, E. (2002) The south to north water diversions in China: Review and comments. *Journal of Environmental Planning and Management,* 45(6), 927–932. doi: 10.1080/0964056022000024415

Cheng, H. & Hu, Y. (2012) Improving China's water resources management for better adaptation to climate change. *Climatic Change,* 112(2), 253–282. doi: 10.1007/s10584-011-0042-8

China Water Risk (2015c) New 'Water Ten Plan' to safeguard China's waters. Retrieved from http://chinawaterrisk.org/notices/new-water-ten-plan-to-safeguard-chinas-waters/

Dillow, C. (2014, August) China's water woes. *PM Network.* 28(8), 8.

Driver, J., Lijmbach, D. & Steen, I. (1999) Why recover phosphorus for recycling, and how?. *Environmental Technology,* 20(7), 651–662.

Fanus, A. A. & Zhao, M. (2012) Impact of irrigation on fertiliser use decision of farmers in China: A case study in Weihe river basin. *Journal of Sustainable Development,* 5(4), 74–82.

Gao, Y., Yu, G., Luo, C. & Zhou, P. (2012) Groundwater nitrogen pollution and assessment of its health risks: a case study of a typical village in rural-urban continuum, China. *PLoS ONE,* 7(4), e33982. doi: 10.1371/journal.pone.0033982

He, J. & Charlet, L. (2013) A review of arsenic presence in China drinking water. *Journal of Hydrology,* 492, 79–88. doi: 10.1016/j.jhydrol.2013.04.007

Hewitt, D. (2015, April 17) China announces ambitious plan to clean up its water, close down polluting factories. *International Business Times.* Retrieved from http://www.ibtimes.com/china-announces-ambitious-plan-clean-its-water-close-down-polluting-factories-1886320

Hu, Y. & Cheng, H. (2013) Water pollution during China's industrial transition. *Environmental Development,* 8, 57–73. doi: 10.1016/j.envdev.2013.06.001

Jin, L., Zhang, G. & Tian, H. (2014) Current state of sewage treatment in China. *Water Research,* 66, 85–98. doi: 10.1016/j.watres.2014.08.014

Jing, L. (2014, May 15) China's South-North Water Diversion Project threatens fish farmers' livelihoods. *South China Morning Post.* Retrieved from http://www.scmp.com/news/china/article/1512107/chinas-south-north-water-diversion-project-threatens-fish-farmers

Kao, E. (2013a, March 18) Hunan villagers refuse to drink tap water deemed safe. *South China Morning Post*. Retrieved from http://www.scmp.com/news/china/article/1193667/hunan-villagers-refuse-drink-tap-water-deemed-safe

Khan, N. (2015, February 1) China is turning fecal sludge into 'black gold'. *Bloomberg Business*. Retrieved from http://www.bloomberg.com/news/articles/2015-02-01/how-the-chinese-are-turning-fecal-sludge-into-black-gold-

Kong, L. (2015, May 4) Inside China's grand plan to fight water pollution. *Market Watch*. Retrieved from http://www.marketwatch.com/story/inside-chinas-grand-plan-for-water-pollution-2015-05-04

Liao, S. (2013, July 9) Water transfer project brings pollution to Northern China. *The Epoch Times*. Retrieved from http://www.theepochtimes.com/n3/165841-water-transfer-project-brings-pollution-to-northern-china/

Li, J. (2013) China gears up to tackle tainted water. *Nature*, 499(7456), 14–15.

Li, J. (2015, February 1) Ways forward from China's waste problem. The Nature of Cities. Retrieved from http://www.thenatureofcities.com/2015/02/01/ways-forward-from-chinas-urban-waste-problem/

Lin, L. (2014a) China's water pollution will be more difficult to fix than its dirty air. *China Dialogue*. Retrieved from https://www.chinadialogue.net/blog/6726-China-s-water-pollution-will-be-more-difficult-to-fix-than-its-dirty-air-/en

Ma, F., Li, C., Wang, X., Yang, Z., Sun, C. & Liang, P. (2014) A Bayesian method for comprehensive water quality evaluation of the Danjiangkou Reservoir water source area, for the middle route of the South-to-North Water Diversion Project in China. *Frontiers of Earth Science*, 8(2), 242–250. doi: 10.1007/s11707–11013–0395–0396

MEP (2012) *China State of the Environment*. Beijing: China Environmental Science Press.

Patton, D. (2015, August 30) China needs to further action to stop water pollution: Vice Premier. *Reuters*. Retrieved from http://www.reuters.com/article/2015/08/30/us-china-water-pollution-idUSKCN0QZ0IA20150830

Qu, W., Zheng, W., Wang, S. & Wang, Y. (2012) China's new national standard for drinking water takes effect. *The Lancet*, 380(9853), e8. doi: 10.1016/S0140-6736(12)61884–4

Tang, C., Yi, Y., Yang, Z. & Cheng, X. (2014) Water pollution risk simulation and prediction in the main canal of the South-to-North Water Transfer Project. *Journal of Hydrology*, 519 (Part B), 2111–2120. doi: 10.1016/j.jhydrol.2014.10.010

Tao, T. & Xin, K. (2014) A sustainable plan for China's drinking water. *Nature*, 511(7511), 527–528.

Wang, H. (2014, May 22) Running the rivers – local government officials are becoming accountable for the water flowing through their jurisdiction. *Beijing Review*. Retrieved from http://www.bjreview.com.cn/nation/txt/2014-05/19/content_619424_2.htm

Wang, M., Webber, M., Finlayson, B. & Barnett, J. (2008) Rural industries and water pollution in China. *Journal of Environmental Management*, 86(4), 648–659.

Wong, S. & Jourdan, A. (2015, April 30) McDonald's fries supplier fined in China for water pollution. *Reuters.* Retrieved from http://www.reuters.com/a rticle/2015/04/30/us-simplot-china-pollution-idUSKBN0NK1BA20150430

Wu, C., Maurer, C., Wang, Y., Xue, S. & Davis, D. L. (1999) Water pollution and human health in China. *Environmental Health Perspectives,* 107(4), 251–256.

Xia, J., Zhang, Y.-Y., Zhan, C. & Ye, A. Z. (2011) Water quality management in China: The case of the Huai river basin. *International Journal of Water Resources Development, 27*(1), 167–180. doi: 10.1080/07900627.2010.531453

Yin, P. (2015, May 21) Solving the water problem: China lays out a blueprint to curb water pollution. (2015, May 21) *Beijing Review.* Retrieved from http://www.bjreview.com.cn/print/txt/2015-05/18/content_688338.htm

Yu, C. (2011) China's water crisis needs more than words. *Nature,* 470(7334), 307.

Zhang, J. (2012) The impact of water quality on health: Evidence from the drinking water infrastructure program in rural China. *Journal of Health Economics,* 31(1), 122–134. doi: 10.1016/j.jhealeco.2011.08.008

Zhang, R., Zeng, F., Liu, W., Zeng, R. J. & Jiang, H. (2014) Precise and eco-nomical dredging model of sediments and its field application: Case study of a river heavily polluted by organic matter, nitrogen, and phosphorus. *Environmental Management,* 53(6), 1119–1131. doi: 10.1007/s00267-014-0268-0

Zhou, Q., Wu, F. & Zhang, Q. (2015) Is irrigation water price an effective leverage for water management? An empirical study in the middle reaches of the Heihe River basin. *Physics and Chemistry of the Earth, Parts A/B/C.* doi: 10.1016/j.pce.2015.09.002

6 Conclusion

This volume has shown that as a result of 30 years of rapid urbanisation and industrialisation, China's water resources have been severely affected by both water quantity shortages and water quality degradation. Groundwater and surface water pollution are major problems in the country as millions of people are drinking unsafe water, and more than half of the country's cities contain water that was termed "poor" or "very poor". It has been acknowledged that water pollution and shortages are far more of a problem in northern China than in southern China. Some of the groundwater in northern China has even been described as "unfit for human contact".

Thousands of industrial facilities are stationed along the major rivers. Polluted wastewater may leach, creating thousands of incidences of what have been termed as water pollution "accidents", or is simply being dumped into rivers from these factories. Industrial facilities are not the only polluters. As reviewed in this book, sources of water pollution include industry (mining, petrochemical industry, chemical factories, manufacturing industries), farming (fertilisers, pesticides, animal waste and antibiotics), and households (faeces and household waste). While nonpoint pollution (e.g. from farms) is difficult to treat, point pollution (e.g. from factories and households) is easier. However, the reality is that the water treatment facilities are insufficient, and there needs to be considerable investment to improve the situation. Even the frequency of water testing is too low, and there are insufficient water-quality monitoring bureaus to assess the level of the problem, so it is not even known what pollutants are in waterbodies. The result is that much of the water, including the groundwater, is polluted. Farmers often have no choice but to use polluted water to water their crops, which may lead to food poisoning and contributes to the country's food insecurity. It is no surprise, considering the discoveries of poisoned and arsenic-contaminated water, that there has been a rise in health problems such as stomach ailments and cancers.

The contamination of surface water and groundwater have resulted in water pollution scandals that led to greater worldwide media coverage and a growing public awareness, at times culminating in strident protests and demonstrations. The Chinese government, despite its lack of transparency, has been forced to address the water crisis plaguing the country. This newfound attention led to a new Action Plan for Water Prevention and Control, and other measures. The regulations are now in place for a drastic improvement of the situation, and the government is willing to invest the necessary funds. Time will tell whether these positive steps will really result in a marked improvement in water quality, or whether provinces will continue to disregard the problem and choose to continue to concentrate on economic growth, as they have done over the last decades.

Bibliography

This is a very concise list of recommended books which provide a more general discussion of the environmental problems in China, in particular those related to water pollution.

Day, K. A. (Ed.) (2005) *China's Environment and the Challenge of Sustainable Development*. New York: ME Sharpe.

Economy, E. C. (2011) *The River Runs Black: The Environmental Challenge to China's Future*. Ithaca: Cornell University Press.

Elvin, M. (2004) *The Retreat of the Elephants: An Environmental History of China*. New Haven: Yale University Press.

Geall, S. (2013) *China and the Environment: the Green Revolution*. London: Zed Books.

Hathaway, M. J. (2013) *Environmental Winds: Making the Global in Southwest China*. Berkeley: University of California Press.

Kelly, W. J. (2014) *The People's Republic of Chemicals*. Los Angeles: Rare Bird Books.

Lora-Wainwright, A. (2013) *Fighting for Breath: Living Morally and Dying of Cancer in a Chinese Village*. Honolulu: University of Hawai'i Press.

Ma, X. & Ortolano, L. (2000) *Environmental Regulation in China: Institutions, Enforcement, and Compliance*. Lanham: Rowman & Littlefield.

Managi, S. & Kaneko, S. (2010) *Chinese Economic Development and the Environment*. Cheltenham: Edward Elgar Publishing.

Marks, R. B. (2012) *China: Its Environment and History*. Lanham: Rowman & Littlefield.

Mertha, A. (2008) *China's Water Warriors: Citizen Action and Policy Change*. Ithaca: Cornell University Press.

Pietz, D. A. (2015) *The Yellow River: The Problem of Water in Modern China*. Cambridge: Harvard University Press

Shapiro, J. (2001) *Mao's War Against Nature: Politics and the Environment in Revolutionary China*. Cambridge: Cambridge University Press.

Shapiro, J. (2012) *China's Environmental Challenges*. Cambridge: Polity Press.

Sze, J. (2014) *Fantasy Islands: Chinese Dreams and Ecological Fears in an Age of Climate Crisis.* Berkeley: University of California Press.

Tilt, B. (2010) *The Struggle for Sustainability in Rural China: Environmental Values and Civil Society.* New York: Columbia University Press.

Tilt, B. (2014) *Dams and Development in China: The Moral Economy of Water and Power.* New York: Columbia University Press.

Watts, J. S. (2010) *When a Billion Chinese Jump: How China will Save Mankind – or Destroy it.* New York: Simon and Schuster.

Zhang, J. Y. & Barr, M. (2013) *Green Politics in China: Environmental Governance and State-Society Relations.* London: Pluto Press.

Index

Notes: Locators in *italics* refer to figures and tables. Ministries are listed under their area of expertise, e.g. Environmental Protection, Ministry of

Printed and bound by CPI Group (UK) Ltd, Croydon, CR0 4YY

17/10/2024

01775686-0005